Philosophy in Science Library

Editors: M. Heller and J. Zycinski

Volume 4
Three Cultures:
Science, the Humanities and Religious Values
by Joseph M. Zycinski

Philosophy in Science Library

General Editors: M. Heller and J. Zycinski

THE WORLD AND THE WORD:
Between Science and Religion
Michael Heller. Translated by A. Chester Kisiel.
For those who have grown up in the positivistic tradition religious faith is impossible not so much as contrary to particular conclusions of science, but rather as having been eliminated by the expansion of scientific methods to all areas of human activity. The author demonstrates that such a position is itself based on an act of irrational faith. He lucidly examines the possibility of faith in the modern world dominated by methods and attainments of sciences.
(Philosophy in Science Library, Volume 1)
ISBN 0-88126-724-4 $ 14.95

THE CATASTROPHIC UNIVERSE:
An Essay in the Philosophy of Cosmology
A. G. Pacholczyk
Professor Pacholczyk presents the historical development of cosmology through a series of catastrophes, with new cosmological outlooks invalidating and replacing earlier ones, rather than through correspondent generalizations, and critically examines recent observations that imply potential new catastropheic changes at the cosmological level in the future. A philosophy of cosmology is considered in the final chapter, emphasizing the limitations inherent in cosmology as a discipline.
(Philosophy in Science Library, Volume 2)
ISBN 0-88126-702-3 $ 9.9

THE STRUCTURE OF THE META-SCIENTIFIC REVOLUTION
An Essay on the Growth of Modern Science
Joseph M. Zycinski
Revolutions occur not only on the level of scientific investigation but also on the level of epistemological and methodological reflection Analyzing transformations in the paradigm of contemporary physics an mathematics the author points out falsification of some methodologica doctrines. The new meta-scientific solutions must not be treated as proclamation of an "absolute relativism" but rather as a new scientifi paradigm free of uncritical dreams of the past.
(Philosophy in Science Library, Volume 3)
ISBN 0-88126-703-1 $14.9

THREE CULTURES: SCIENCE, THE HUMANITIES AND RELIGIOUS VALUES

Joseph M. Zycinski

Pachart Publishing House

Tucson

Joseph M. Zycinski (1948-)
Three Cultures:
Science, the Humanities and Religious Values

Library of Congres Catalog Card Number: 87-062402
International Standard Book Number: 0-88126-727-9

Philosophy in Science Library Volume 4

Pachart Publishing House
A Division of Pachart Foundation
A Nonprofit Association
1130 San Lucas Circle
P.O. Box 35549
Tucson, Arizona 85704

Contents

Introduction 7

1. The Breakdown of Ancient Order 13

2. Subtle Spirit and the Machine of the World 23

3. Deep Dimensions of Scientific Evolution 29

4. Between Mathematics and Mysticism 35

5. God Hidden in Nature 41

6. Rationality and the Drama of Golgotha 47

7. God Omnipresent Incognito 53

8. Waiting for a New Ferguson? 59

9. The Closed Society and Its Religious Enemies 69

10. In Defense of Scientific Reason 79

11. Love of Life or Fondness for Beauty? 87

Notes 93

Introduction

The most recent period of intense development in science and technology has led to fundamental revisions concerning the appraisal of the socio-cultural consequences of science. At the beginning of our century science was still treated as a substitute for religion for the enlightened and a source of cultural transformations forming a new mankind. Henry Adams, in describing his feelings about the Paris World's Fair in 1900 professed that he felt in the *Galerie des machines* the same as the first Christians must have felt at the foot of the cross. The fascinating world of technology enchanted him by its new perspectives and it provided reasons for expecting a new society and culture of the future. In this delightful perspective even the ordinary dynamo appeared as an unusual symbol, as a kind of "moral force" (1) and preview of a brave new world.

Sentiments similar to those of Adams were held independently of cultural and political differences. Signs of the same enthusiasm, hopes and optimistic expectations emerged even in centers separated by barriers of space and political systems. Francis Klingender convincingly illustrates how much the technological enthusiasm of the Russians, during the building of the Dnieprostroj Dam, caused one to think of comments of the British press which 100 years earlier praised the achievements of the Industrial Revolution.(2)

Gradually the earlier enthusiasm turned into skepticism. Science, which in the 19th century was meant to be a source of universal happiness, today is presented more often as a source of current evil that deprives human existence of its natural beauty. The previous uncritical praise of science gives way today to an equally uncritical denial of the value of science. Slogans that extol the return to nature, despite their superficial references to Jean Jacob Rousseau, very often become mere labels for an unsophisticated anti-intellectualism. The ideal of the human person liberated from theoretical reflection and from the great metaphysical questions is shown as a new pattern of existence for creatures who reached a stage of disappointment in science and who have recognized the bitter taste of its technological results.

In contemporary critiques, the so-called scientific culture is opposed to a humanistic one, in the same way that in the 19th century science was opposed to religious beliefs, theology and metaphysics. Both of these oppositions are substantially without foundations. Their arbitrariness is acknowledged by representatives of post-positivist science as well as by former critics of religion who have discovered the real value of the subject of their critiques. Abraham Maslow, who belongs to the latter group, suggestively portrays his path away from religion conceived superficially as a set of rituals and practices to a religion in which are explored profound questions of the ultimate meaning of the universe as well as of the meaning of human life. When naïve anthropomorphisms are overcome, one discovers the search for perfection and the adherence to spiritual values as "the essence of the religious tradition." This discovery discloses the humanistic dimension of our religious attitude. It facilitates human self-actualization and creates a chance for "a great flowering of a new kind of civilization." (3) In the religious tradition understood in this way there disappear the former oppositions that have been propagated by the partisans of the philosophy of positivism. Social harmony, spiritual peace and intellectual efforts become constitutive elements of human self-actualization that are founded upon our essential openness to the transcendent divine reality.

The special obligation of looking for a new cultural synthesis rests on the representatives of what has been called Christian thought. Modern science originated and developed in the sphere of the intellectual and cultural influences of Judeo-Christian religion. We recognize the year 1687, the year of the publication of Newton's *Principia*, as a symbolic beginning of modern science. Consistently, we may, and we must, formulate the question: What is the mutual relationship between modern science and religious thought after 300 years of their coexistence?

The answer to this question do not provide special reason for optimism. Very often the representatives of religion treat science as a source of interpretative difficulties or as a cause of dramatic world-view conflicts. Relatively often, new scientific theories have still been treated with distrust because they bring up embarassing questions and destroy the simplicity of well-ordered religious beliefs. It is easier to appreciate the formal order than the objective truth. Such an attitude, however, is strikingly incompatible with the words of Christ:"the truth will set you free." (John, 8:32) The deliverance brought by the Christian truth requires the ability of departing from a certain style of ordered thinking in which simple stereotypes are appreciated more than is the effort of creative reflection that aimed as setting us free.

In these essays, when I speak of religious values I mean primarily values that constitute the Judeo-Christian tradition in whose influence modern science was born and where, therefore, we should look for the intellectual premises that would facilitate a new synthesis of the cultures which are now separated. The arguments that I propose here relate, nevertheless, to many other religious traditions. They remain valid if, following Maslow, we understand religious values as the values that, above and along with the self-actualization that is dependent upon our openness, facilitate discovery of the integrating divine sacrum.

Intellectual stereotypes which previously caused mutual distrust between science and religious faith contributed to a simplistic image of science that was popular among the adherents of scientism and positivism. Their followers believed that the whole truth about reality is discovered in the natural sciences, while metaphysics and theology contain merely impressive pseudo-explanations. This faith in the cognitive monopoly of science found a special expression in the *Manifesto* of the Vienna Circle that was published in 1929. This *Manifesto* ensured that the whole of reality can be described in sentences of the language of observation, because in science there is no place for speculations concerning the hidden unobservable depths. In contradistinction to such tenets, contemporary philosophy of science warns against a programmatic avoidance of "the unobservable depths." Because of the acceptance of the anti-depth program a scientist can lose a trip to Stockholm for the Nobel Prize. (4)

The hidden structures that are not accessible to immediate observation constitute the basic subject of interest for modern science. Contemporary discussions concerning Grand Unified Theories in physics, vacuum fluctuations in quantum electrodynamics or deep structures in linguistics focus upon the search for a hidden fundamental structures which fall outside the sphere of our direct observations. The new intellectual climate that reigns in post-positivist science creates an exceptional opportunity for an effective searching for the affinity between the areas of scientific and religious truth.

It is worth noting in the context of the development of modern science that theological premises have often played a positive heuristic role in the growth of new scientific theories. Religious belief in the order of nature established by God inspired Kepler's new astronomy and Galileo's mathematical physics. Both these authors believed that God reveals His wisdom in simple mathematical equations as well as in the stability of physical laws. In their approach, scientific reflection inevitably had a theological dimension because it is a means of discovering God's logos immanent in the cosmic order. When, in the process of the growth of modern science, scientists' attention was focused upon the technical aspects of the research that was undertaken, the classical doctrine of God's immanence in nature became relatively forgotten. The critical development of this doctrine as well as its accommodation to the thesis of divine transcendence toward nature creates an exceptional opportunity for the unification of distant branches of science, human values and religious truth.

In attempting to sketch in this book a perspective for such a unification, I call upon the philosophical principles accepted by Alfred North

Whitehead and Charles Hartshorne. The most fundamental theses that disclose the theological dimensions of science can be, nevertheless, formulated and justified independently of the cognitive framework of Whiteheadian process philosophy. I think, however, that this philosophy, because it combines reflection upon the results of scientific discoveries with the experience of the poetical aspects of nature, creates an exceptional occasion for philosophical dialogue between religion and science. In this very perspective, the human pursuit of truth and beauty becomes an element of great intellectual adventure which destroys the barriers of the previous divisions and belongs to the essence of civilization. (5)

I would like to express my gratitude to Carol Suplicki, Florence Hetzler, Albert Wojtcuk and Michael Winters for their valuable suggestions and inspiring comments.

1

The Breakdown of Ancient Order

> I am Fire, and Ayre, my other Elements
> I give to baser life.
> Shakespeare, *Anthony and Cleopatra*

In contemporary poetry it is rather difficult to find expressions of romantic feelings mixed with scientific references to the chemical composition of the bodies of our beloved ones. Contrary to present custom, Shakespearian *Cleopatra* expresses in poetic form the physical knowledge common to the epoch when the play was written. According to Aristotelian physics that still dominated in the 16th century, all people were composed of four elements: fire and air, water and earth. In the moment of death, the elements of fire and air were to leave the human body and to be carried to heavenly regions that consist of subtle aethereal matter. The "other Elements", water and earth, were to remain in the sublunar region marked by decay, variance and imperfection.

The mystery of human life and death had in such physics evident cosmic consequences. The ordered universe provided patterns of moral goodness and aesthetic beauty. The emptiness of cosmic spaces did not horrify the admirers of Aristotelian physics because the whole universe was thought of as a single organism which can be described in well-known categories of everyday life. The earth-centered cosmos seemed to be a fatherland of mankind, since its structures appeared to be immediately related to human existence. This friendly world of medieval cosmology was supposed to be animated by the *anima mundi*, the cosmic soul analogous to human souls. It was William Gilbert who argued in the 16th century that calling into question the existence of the world's soul implies a degradation of the universe, while regarding the world as a mechanism insults its Creator.

The Scientific Revolution originated by Copernicus and Galileo provided not only new scholarly standards of research but also new insight into the mutual relationships between the human, religious and scientific cultures. When Copernicus worked on *De revolutionibus*, he was convinced that scientific and religious truths are consistent and argued that it is philosopher's duty "to seek the truth in all things, to the extent permitted to human reason by God." In his Preface to the *On the Revolutions*, the Polish astronomer expressed his belief in the basic unity of scholarly and religious cultures when he addressed the Pope and informed him: "in this very remote corner of the earth where I live, you are considered the highest authority by virtue of the loftiness of your office and your love for all literature and astronomy..." (1)

The dramatic events that followed the condemnation of Copernican theory in the process of Galileo resulted in both a new hierarchy of authorities and new attitudes of theologians toward science. The deplorable sentence of 1633 should not be thought of as a result of internal conflict between theology and science, but merely as a socially conditioned event that disclosed the supremacy of canon law over theological reflection in the post-Tridentine Church. (2)

The affair of Galileo can hardly be approached as a conflict between theology and modern science, because the new science presupposed important theological tenets. Both in his *Dialogue* and in *The Assayer* Galileo contended that the mathematical description of physical processes is possible because of the existence of the Divine Geometrician who determined the nature of physical phenomena. This vision of mathematics grounded in God is defended by Salviati, the port-parole of Galilean views, when in the *Dialogue* he follows Plato in arguing that the human intellect "is to participate in divinity" on account of its grasp of mathematical knowledge. (3) The book of Scripture and the book of natural revelation are complementary. The experimental study of nature is necessary for our intellectual development because, as it was argued by Cardinal Baronius, the intention of the

Holy Spirit is to teach us how one goes to heaven, not how the heavens go. Galileo repeats the Cardinal's statement in his letter to the Grand Duchess Christina of Lorraine. He goes back even further to the teaching of the early Church to quote Tertullian who argued: "that God is known first through Nature, and then again more particularly by doctrine; by Nature in his works and by doctrine in his revealed word."

After accepting such principles Galileo points out internal inconsistencies in the standpoint of these authors, who attempt to justify theologically their lack of confidence in modern science. Such a standpoint cannot be consistent, since on the one hand its adherents believe that God has endowed us with senses and intellect, while on the other hand they hold that the use of our cognitive powers should be suppressed. Rejecting similar inconsistency, Galileo defends the thesis of the essential unity between scientific and religious truth. This unity stems from the fact that "the Holy Bible and the phenomena of nature proceed alike from the divine Word," i.e., the Logos that is present both in the Gospel and in the mathematical conjectures of the Pythagoreans. There are no reasons to mourn over the Aristotelian model of the cosmos as an organism. Its theological significance seems ambiguous, at least of those versions which located the throne of Lucifer in the center of earth. Such an approach resulted in the model of a devil-centered universe, and its apology can hardly be based on theological premises. Only a lack of creative imagination could have facilitated its categorical defense that nourished distrust toward new routes of human intellectual pursuit.

The opposition against the new direction of scientific thought stemed not only from religious traditionalism but also from cultural humanism, the doctrine very influential in the time of the Renaissance. John Donne was not alone in his complaints when he lamented that the "new philosophy calls all in doubt" and eliminates even the element of fire from the cosmological picture of the world. Paintings that presented Copernicus in hell, chained to the throne of Lucifer, provide adequate testimony to the cultural protests against the scientific revolution which called into question the central position of Man in the universe. A *Cypress Grove*, published in 1623 by William Drummond, furnishes gloomy descriptions of the sickness unto death induced by rejection of the Aristotelian order and acceptance of the new science. "The new philosophy" destroyed the cosy universe in which everything was described in anthropomorphic terms akin to our natural intuitions. The image of volcanos erupting with anger and the picture of an alive earth giving birth to metals were replaced by an abstract mathematical description of the world, the world hard, cold, colorless, silent and dead.

It is interesting that this shocking new picture of nature received an approval in so short a period of time. At the beginning of the 18th century, in Alexander Pope's poems one can find a new vision of human culture based upon Newtonian physics. At the end of the same century, in 1794, two poets, Johann Goethe and Friedrich Schiller, met each other during a session of the Jena Scientific Society. The new alliance between modern science and humanist culture was particularly attractive for those humanists who did not try to understand the new physics, but were deeply satisfied with quoting Pope's compliments to Newton, which said that after Newton everything became clear end evident.(4)

As a matter of fact there were many problems unclear in Newton and after Newton. Physicists as well as philosophers discussed them in detail. Even Goethe, like the last Mohican among the poets, tried to humanize the theory of optics by rejecting the views Newton had expounded in the *Opticks*. In the new cultural and intellectual climate, it was much easier to express one's admiration for modern science than to try to understand it. When Richard Bentley worked on his sermons in which Newton's *Principia* was to be discussed theologically, he consulted the mathematician John Craig, asking him what books should be read to understand the work. In response he received a bibliography consisting of 40 publications which were to elucidate the mathematical jargon used by Newton in his exposition of the tenets of the new physics. Bentley, a philologist, despaired after receiving the bibliography; in the short time that he had at his disposal it was impossible to read the 40 books Craig had recommended.

Regardless of the fact that the objective value of Bentley's sermons seems at least dubious, his attempts to create a new synthesis of scientific, philosophical and theological truth inspired a new style of intellectual openness to novel scientific theories and brought creative revisions to the traditional image of the world. This style was particularly fostered by Robert Boyle, the English chemist, physicist and theologian. When dying in 1691 Boyle expressed in his will that each year a series of sermons should be preached to interpret new scientific discoveries from the standpoint of Christian faith. Some historians repeat the witty comment that nobody thought seriously about a conflict between religion and modern science before Boyle suggested reconciling them. The overactive search for scientific confirmation of religious beliefs led to the skeptical denial of earlier theological interpretations and to the acceptance of the philosophy of deism, then in vogue. The main problem for theology arose, however, not from overrating new scientific theories but rather from ignoring them and combining theological insights with the outdated Aristotelian cosmology.

The present isolation between scientific, religious and humanistic cultures was by no means created by scientists. The founders of modern science were, in general, sensitive both to the metaphysical presuppositions of the adopted methods of research and to the social-cultural effects of their discoveries. Those who professed intellectual traditionalism could have been found much more easily among the representatives of art and philosophy who endeavored to enlighten immature science by determining its moral and aesthetic goals. Sharing such an approach, many followers of Goethe criticized Newtonian optics by repeating arguments that the artificial conditions of laboratory observations destroy the natural beauty of light and strip physical phenomena of their natural charm. Similar objections presupposed that the new physics should contain the same moral and existential ingredients which belonged to medieval science. To justify this supposition it was not enough to quote authorities of the past and defend obsolete philosophical cosmology.

Consequences of this attitude are manifest in various philosophical and theological works published in the 18th century. Their authors, after quoting time-honored authorities, felt relieved of any duty to critical reflection. They disguised their lack of elementary concern in the new scientific discoveries by introducing pathetic declarations which announced that their works were written according to modern academic standards. C.R. Billuart's *Summa S. Thomae hodiernis academiarum moribus accomodata* can serve as an example of these standards. In it no mention is made of Copernicus or Galileo. The empyreal heavens are described as the subject of research of 18th century cosmology, and the motion of planets is explained by charging the angels with the duty of controlling planetary orbits. Such a mixture of angelology with kinematics manifests the deplorable hibernation of thought in the centers that were supposed to study the perennial truth and examine the structure and the order of the universe. In this framework, intellectual conservatism became a virtue as well as a hallmark of lofty indifference to the gossip-mongering which allegedly was practised in the natural sciences.

As a result of these attitudes, philosophical commentary on the new science was developed by those 19th century authors who were not familiar enough with either science or philosophy but nevertheless assumed the role of spokesmen for progress and ultimate truth. Their private visions, presented as scientifically grounded prognoses, yielded a new picture of a happy mankind that is disentangled from both traditional metaphysics and religious beliefs. Modern science which arose in the milieu of Judeo-Christian culture appeared in the shockingly new role of the social-cultural opponent of any religion. Both the mutual lack of confidence between scientists and other intellectuals and the conflicts which often marked their relationships led to important consequences on the level of cultural life. The Renaissance ideal of integral human culture turned out to be unrealistic in an epoch facing the coexistence of isolated subcultures.

The present painful isolation is regarded by certain members of religious communities as a form of splendid and inevitable isolation. They oppose the perennial truth of religious doctrine is opposed by them to the ephemeral fluctuations of illusive scientific conjecture. Such an approach expresses nothing but a philosophy of easy resignation. One ignores religious principles that define our attitude toward God Who can be recognized in rational analysis of natural phenomena -- an attitude which constitutes the very essence of religious insight taught in both the Old and the New Testament and which characterizes the practice of the early Church as well as the contemplative insight of St. Francis of Assisi.

A consistent and well-considered theological justification of Christian concern for natural science may be found in treatises worked out by the influential School of Chartres in the 12th century. It was William of Conches who categorically denied attempts to oppose theology and the study of nature.In his arguments, God the Creator of the human intellect and of the order of nature is to be glorified through our understanding of the natural order. Theological recognition of the importance of physical regularities was based on the Vulgate translation of the Epistle to the Romans (13:1): *quae a Deo sunt, ordinata sunt* whatever is from God, is ordered. For many reasons the Platonic picture of the cosmic order described in the *Timaeus* appealed to the imagination of theologians who belonged to the School of Chartres and inspired their works on the harmony of natural and supernatural values. The beauty of this harmony was impressively depicted by Alain of Lille. In his description of nature, which is subordinate to the divine Mind that brings order out of chaos, we find the anticipation of both Blake's poetry of nature and Francis' cosmic mysticism.

In the religious hymns of the School, fascinating descriptions of nature unite both physical and theological perspectives. This integration does not yield uncritical physical theology, but instead provides a beautiful description of cosmic harmony that embraces aesthetic, moral and existential values.(5) Germs and light, human love and physical laws are united together in Alain's philosophy of nature. His philosophical and physical descriptions disclose the theological dimension of nature presenting the physical world as the domain of the presence of the immanent God.

The image of nature proposed by the Chartres School met violent opposition from some medieval scholars. William of Saint-Thierry, for instance, attacked the very idea of uniting biblical doctrine with the pagan philosophy of the *Timaeus*. In spite of the categorical protests of William's followers, the School provided intellectual patterns in which harmonious synthesis replaced the isolation of various cultural traditions. With these patterns of new harmony, the world of natural phenomena was no longer an agent of alienation. It appeared, on the contrary, according to Bernard Palissy's later comment, as "the beautiful book of heaven and earth offered to every man."

The importance of reading this "beautiful book" and contemplating its beauty was recognized by many other medieval authors. They continued the intellectual tradition of St. Augustine for whom the necessity contained in mathematical truth revealed God's existence and led directly to the rejection of atheism. In his view there was no rivalry between natural and supernatural reality; theology competed neither with philosophy nor with science. Revealed truth and rational argument, sense experience and contemplation of nature were put together to constitute the harmonious ideal of human existence rooted in a cosmic horizon. In a similar manner, St. Thomas Aquinas defended the essential compatibility of natural and supernatural truths. In his *Summa contra Gentiles* Aquinas argued that the investigation of created beings appears both useful for understanding religious truth and necessary for a better knowledge of God. The same stance may be found in Duns Scotus' *Reportata Parisiensia* where he defends an essential correspondence between human rational wisdom and the wisdom that was revealed in Christ.

This belief in the harmonious coexistence of scientific, religious and moral truths was basic to the investigations of the founders of modern science. Accustomed as we are to the dry and colorless style of contemporary science, we are amazed when in Kepler's geometric description of the solar system we encounter an expression of his religious and aesthetic feelings such as the text that describes God "who ... sent birds supported on wings into the wilderness of the air; who gave white bears and white wolves to the snowy regions of the North...; who gave lions to the deserts of burning Libya and camels to the wide-spread plains of Syria... ." (6)

To yearn for a return to the Keplerian style in modern science would be futile. The rigid methodological principles underlying the foundations of natural sciences definitely exclude from their domain of science any appeal to extra-scientific or personal aspects of one's research. No scientist would seriously consider the work of philosophers who would try to rehabilitate medieval standards of knowledge. According to the prevailing paradigm of scientific practice, the results of an investigation are considered values much more important than their social-cultural consequences. In our present scientific-technological civilization, we must employ new means to discover the original harmony that was destroyed after tasting the tantalizing fruits of the tree of scientific wisdom.

2

Subtle Spirit and the Machine of the World

The Galileo-Newton scientific revolution brought to an end the traditional cosmology in which angels were to take care of celestial motions. To be sure, there were many 17th century authors who expressed doubts with regard to the role of angels in celestial dynamics. But they did not always stem from the principles of physics; sometimes they depended on outdated physico-theology which contained naive anthropomorphisms. It was Leonardus Lessius who at the time of the principal discoveries of Galileo explored the issue of whether or not finite and imperfect angels are creatures responsible and competent enough to supervise the heavenly motions. Other cautiously authors tried to avoid similar speculations and confined their research to the problem of God's transcendence. They evaded the troublesome questions posed by the new physics by vaguely referring to the empyrean. Religious reflection on nature appeared more often in the writings of physicists and astronomers than in those of theologians.

A risky mélange of new science and fantastic philosophical speculations may be found in the visionary mysticism of Jacob Boehme or in the cosmological poetry of Campanella. Such an uncritical mixture of astronomy, theology and poetry is already seen in the titles of works like *Astro-theology* or a *Demonstration of the Being and Attributes of God from a Survey of the Heavens*. The good intentions of their authors were above the logic of argument. In contrast those authors who relied on solid arguments more than on speculation did not see a need for introducing poetic vision into the new physics, especially when their attention was focused upon the new type of poetry which was contained in the mathematical formulae of Newton's dynamics.

In this new intellectual climate, only a few authors offered non-trivial reflections concerning God's immanence in nature. Among the most original of these was Baruch Spinoza, a Dutch Jew from Amsterdam. Introducing the formula "God or Nature" (Deus sive Natura) he left himself open to the criticism that he identified God with nature. It is difficult to determine which of his formulations were dictated by his diplomatic caution and which by his rhetoric. Nevertheless, in contemporary debates on Spinoza's philosophy one to demonstrates that the charge of pantheism is baseless because many attributes he ascribed to God cannot be related to nature (1). It is argued that Spinoza's alleged pantheism stems more from a simplified identification of God with nature in the laconic slogan than from an intimate analysis of the philosophical doctrine expounded in the *Ethics*. Setting aside this question, we cannot deny that for a long time the fear of being accused of pantheism was a basic factor inclining many philosophers to avoid risky questions about God's immanence in nature. It was better and safer to focus theological reflection upon God's transcendence.

The painful experiences connected with the trial of Galileo and the condemnation of Spinoza had their consequences not only in philosophy and natural theology, but also in art. Owing to diplomatic caution, many 18th century painters prefered to avoid depicting nature. The same beauty of nature that was earlier acknowledged as a sign of God's presence in the world suddenly became an element raising the suspicion that the given painter shared Spinoza's pantheism. Voluntarily chosen silence, in philosophy as well as in art, were meant to guarantee religious orthodoxy. They also effectively guaranteed neglect of key philosophical problems.

Just as Spinoza involuntarily discredited the term "pantheism", Richard Bentley brought "atheism" into disrepute. In 1693 he delivered a series of sermons containing theological interpretation of Newton's physics. These sermons were received with great public respect out of proportion with their true value. In fact the published version of his sermons had many more readers than Newton's *Principia*. Their lengthy title suggested that atheists could be found only among these people who never studied the principles of the new physics. As a result, 18th century adherents of atheism called themselves deists, to avoid possible suspicions about their scientific ignorance.

It is important to recognize that deists and agnostics at the turn of the 18th century did not attempt to justify their rejection of a religious world view from a scientific basis. The reasons for their views were rather connected with their hesitation with regard to religious institutions or with personal doubts concerning the existence of hell, the doctrine of divine Providence, and the literal interpretation of the Bible. Skepticism concerning such matters was supposed to be proof of one's intellectual independence. It was not a conclusion based on arguments directly depending on the new scientific theories. The God of deism was a creator who had no interest in his creation. He was to remain remote from the situations of human suffering and existential dilemmas. Such a conception of God had no foundation in either Newton's philosophy or his physics.On the contrary, Sir Isaac in his *Opticks* as well as in letters to Bentley emphasized that the immanent God reveals his presence in the stable laws of nature and in the mathematical harmony of natural phenomena.

Many times Newton attempted to solve the disturbing problem of whether or not the phenomena of gravity and of the cosmic ether were signs of the presence of a subtle divine spirit who invisibly fills the whole of space. In the first edition of the *Principia* Newton carefully avoided these problems for, as he explained in the second edition, he did not want to multiply untestable hypotheses. In his correspondence, however, he consistently returned to the question of whether or not the harmony of the cosmos expressed in mathematical formulae discloses an element of invisible divinity present in the physical world. One hundred years earlier the same question intrigued Kepler when in the elliptic shape of planetary orbits he perceived the immanent logos impelling one to reflect upon the mysterious harmony of nature.

One century after the death of Newton there was a new and different style of scientific reflection. Classical metaphysics became a symbol of intellectual obscurantism. The partisans of Comtean positivism categorically rejected the Platonic-Pythagorean tradition which had been so important in the rise of modern science. In an age caught up in technological achievement, Kepler's mysticism of nature seemed merely a useless burden and a relic of the pre-scientific stage in human intellectual development. Such an atmosphere facilitated the integration of Newton's physics with the philosophy of mechanism, in which God was only a dispensable hypothesis. Scientists and engineers fascinated by the functioning of the world machine were no longer interested in discussions about a transcendent God. Theologians, in turn, discouraged by groundless critiques developed by the representatives of naïve materialism, treated science with distrust. As a result, a deep division arose between the truths of religion and the truths of science.

Although the philosophy of mechanism collapsed at the end of the 19th century and the self-appointed spokesmen of science became recognized merely as eloquent propagandists, certain stereotypes of mechanistic thinking were more persistent. In the context of the Einstein-Planck revolution, dogmatic support for these stereotypes appears as a sign of intellectual inertia. The poetry of nature discredited by yesterday's positivism is today reaffirmed by the representatives of the so-called Princeton neognosis (2). In their writings one can find both amateurish philosophy and quasi-religious mysticism mixed with scientific data in order to construct a consistent world view, free of gaps that resulted from excluding metaphysical questions from our intellectual concerns. Many contemporary attempts at new synthesis of philosophy and science are open to legitimate critique; their openness to metaphysical dimensions of scientific research should be recognized as a correct reaction to naïve schemes that were developed when positivism dominated the philosophy of science.

In our present post-positivistic periode, when the universe is no longer envisaged as a world-machine but rather as a cosmic brain or organism (3), new philosophical perspectives are required to critically elaborate the doctrine of God's presence in nature. This task cannot be fulfilled if the so-called natural theology does not become more natural, that is, more open to the results of the natural sciences.

3

Deep Dimensions of Scientific Evolution

> The self-confidence of the learned people is the comic tragedy of civilization.
> A.N. Whitehead, *Immorality*

A great many painful phenomena of contemporary civilization have their intellectual roots in the optimistic self-confidence of the learned people of the 19th century. Significant discoveries made in that century in the fields of biology, chemistry and physics led to the development of visionary philosophies. Although the scientists themselves maintained considerable distance to philosophical comments dealing with their discoveries, the very results of these discoveries became a favorite subject of discussions among the enthusiastic boosters of science. In the texts of the latter an uncomplicated equation appeared, stating that science = progress = happiness of mankind.

Different versions of similar philosophies stressed the importance of different details in their approach to the imagined perfect world of the future. August Comte's positivistic futurology differed essentially from Karl Marx's prognoses concerning a classless society, although both visions were inspired by the belief in the inevitability of progress which was supposed to bring a happy end to the infant disease of human society.

The optimistic faith in progress had been proclaimed by philosophers and scientists. Many supported Ernst Haeckel's opinion that it was through science that all enigmas rankling mankind from time immemorial had finally been solved.The same confidence in accomplishments of 19th century science was shared by those physicists who alleged physics to be on the point of reaching the limit of its discoveries, which would consequently doom future generations of physicists to boredom, their main task being, at the very most, a more accurate calculation of physical constants. The optimism of the epoch was expressed by the conviction that, in other fields of science, beyond physics, a new host of Newtons and Darwins was just about to appear. It was not without reason that Marx was pronounced by Engels the Darwin of social sciences. Marxian analysis concerning the evolution of human societies was meant to show the inevitable mechanisms of development in which class struggle was the counterpart of the evolutionary struggle for survival.

At the turn of the 19th century, the optimistic faith in breakthrough discoveries manifested itself in Freud's attempt to define the mechanisms controlling human subconsciousness. The attempt, soon regarded by Freud himself as a symptom of another Copernican revolution, today provokes only bitter comments of the authors who defend rational principles in scientific explanations. Ignoring these principles in his revolutionary approach, Freud felt entitled, for instance, to formulate opinions on woman's nature in general, despite the fact that his empirical data were limited to his docile wife and a group of neurotic woman-patients who regarded him as the supreme authority. His ungrounded generalizations and arbitrary conjectures, presented as ultimate truths, provided methodology with standards that still shock not only feminists who reject Freud's model of a woman with her many complexes, but also the majority of critical authors.

A comparison of procedures applied bv Newton with those adopted by the 19th century propagators of science leads to rather sad reflections. Working out his theory of gravitation, Newton put forward six different hypotheses to explain the phenomenon of gravity. In the course of time he compared their explanatory values and examined their empirical confirmation. The 19th century enthusiasts of science were not bothered by such problems. Their prognoses were meant to provide both the only just and necessary solutions, and the ultimate criterion of their apparent correctness was the optimistic faith in the exceptional character of the epoch.

The Einstein-Planck revolution in physics diverted the scholars' attention to totally new questions about which the 19th century experts in science and progress were completely ignorant. In the context of new discoveries, particularly when reading texts and prognoses more than 100 years old, we can learn intellectual humility. We are taught a meaningful lesson observing on the one hand boisterous slogans expressing common-sense wisdom of the epoch, and on the other hand, a complete silence about many phenomena characteristic of our epoch. In the atmosphere of Darwin's theory of natural selection, it was easy to develop futuristic visions in which 20th century strawberries were supposed to be the size of pumpkins; and all that due to the effort of inestimable biologists. Strangely enough, the same visions did not predict the appearance of computers, though automatization of labor resulted in consequences important not only for Marx's economic forecasts.

In spite of clear faults with the old faith in progress-bearing science, it cannot be denied that this faith also fulfilled positive functions. Its initial outcome was the cultural integration in which both the scientific discoveries and the attempts at new analysis of great metaphysical questions were the field of interest. Even though 19th century propagators of science can be blamed for spreading the humanistic propaganda of science, there is no doubt that this sort of propaganda prevented the phenomenon of two cultures, a painful split between the scientific and the humanistic culture. The problem of this cultural split was already examined in a number of monographic studies, among which C.P. Snow's *Two Cultures*(1) maintains a classical position.

Since the first edition of this famous work, not only has the gap between the humanistic and scientific culture widened, but also the scientific culture has undergone some changes and developed several independent subcultures. The phenomenon, often termed as departmentalization, or Balkanization, of science (2), leads to various negative consequences. Within the same Institute of Physics, the cosmologists use a language every day less and less understandable to the experimental physicists. A specialist in sedimentary rocks feels frustrated and lonely at a symposium on igneous rocks. A representative of any field of science who has not yet mastered computer programming feels like Australian aborigines at the Greek Areopagus when his fellow scientists discuss the subtleties of software. A result of such situations is the atomization of science which manifests itself in the lack of understanding for the interdependence of different branches of science and the lack of its comprehensive picture. Nothing is said about essential differences in the meaning of the word "science" which for the representatives of the demistyfying social sciences or psychoanalysis is not the same as for specialists in theoretical physics.(3)

All this leads to a conclusion that science is becoming less and less understandable to average newspaper readers, which in turn makes them look for its substitutes in occultism, astrology or parascience. Specialistic academic studies may very easily become self-consistent and self-referential. When this happens we can expect that professional publications in the field of sociology of science will be clear exclusively to the sociologists of science who will reach out for them when supervising doctoral dissertations in sociology of science.

It is not my intention to define in these essays the concept of culture. Regardless of the adopted definition of culture, the sad phenomenon described by Snow as the bifurcation of the scientific and the humanistic culture has a no less painful counterpart in the split between the scientific culture and the religious thought. This split did not exist when Copernicus addressed in his *De revolutionibus* Paul III as "the highest authority," not only because of the position held but also because of his love for literature "and astronomy too."(4) Dramatic consequences accompanying the growth of modern science resulted in the course of time in the change of the status of former authories as well as the appearence of many prejudices concerning the relationship between religion and science. The painful experience connected with Galileo's astronomy and Darwin's theory of natural selection made theologians reluctant to refer to achievements of the natural sciences. Barriers of different languages, mentality and ways of interpreting the world only helped to keep up the silencing iron curtain between the fields of science and religion.

The present situation is deeply rooted in history, but its grotesque character cannot be fully justified by historical reasons. Particularly, when in the new intellectual climate representatives of physics make, for instance, such shocking statements: "in my opinion science offers a surer path to God than religion."(5) Apart from the fact whether one accepts or not P.C.W. Davies' opinion as quoted above, one cannot deny its existence and choose to be dumb on the subject of the relationship between religious faith and scientific discoveries. The consequences of such choice would be not less painful than those resulting from the separation of the humanistic and the scientific culture

Modern science has originated and developed in the sphere of cultural influence of the Judeo-Christian thought; any attempts to isolate it from this cultural milieu are bound to meet various objections. Unification is not only the dream of physicists working on Grand Unified Theories. A unified picture of the world in which scientific, human and religious values are harmonized with each other is a subject of great intellectual interest for every person who wishes to rationally define her or his attitude towards the world. A difficult task of the unification of three cultures -- scientific, humanistic, and religious -- could be carried out with the help of the language of such philosophy that would be open to the development and achievements of the natural sciences. In my opinion, a philosophical system particularly suitable for reflections on the results of contemporary science is the metaphysics of process as found in the version presented by Alfred N. Whitehead in his *Process and Reality*. This philosophy undertakes the great questions that continue to intrigue mankind since the pre-Socratics. Because of its recognition of both scientific data and the aesthetic aspect of the cosmic order, it creates an auspicious opportunity to counteract the phenomenon of three cultures.

To underestimate the role of philosophy in the process of cultural unification seemed natural during the period dominated by positivism. In this doctrine philosophy was meant to fight classical philosophy. Metaphysics was regarded as a form of poetry or an expression of intellectual immaturity of societies cultivating metaphysical problems. Such approach exerted significant influence on the mentality of many physicists who were about to gargle in their mouths after pronouncing the word "metaphysics." The death of logical positivism in the early 1950s facilitated a parting from many simplifying schemes formerly imposed by positivism. A change in approach to the problem of relationship between science and philosophy is manifested, for instance, by the increasing number of publications which contain scientists' comments on philosophical aspects of new scientific discoveries. In the past, metaphysics was likely to be treated as an occupation for elderly men who loved poetry and had already nothing more to say in science. A relic of that epoch may be found in the adage: "First sclerosis, then synthesis." Its content indicates that even the elementary questions about synthesis and the unity of the world view seemed then suspiciously close to allegedly meaningless metaphysical reflections.

Today philosophical interests appear as evidence of opening to deeper truth. They are to testify that science cannot be cultivated as a craftsman's trade but has to discern the hidden meaning and mutual links between truths discovered in different domains. In such situation, neglecting the questions concerning the inter-relationship of the three cultures would be an inadmissible evasion. "Philosophy is mystical. For mysticism is direct insight into depth as yet unspoken" wrote Whitehead at the end of his *Modes Of Thought*. The same element of the insight into depth appears in different versions in the religious experience, in philosophical debates about arche, in the reflections of a theoretical physicist which all go beyond the measurable world of the immediately given.

4

Between Mathematics and Mysticism

> Il est possible que le progrès ne soit que le développement d'une erreur.
>
> Jean Cocteau

In the beginning of the 19th century, Wilhelm Hegel in his *Encyclopedia of Philosophical Sciences* expressed his conviction that stars and material structures of the world do not constitute for philosophers a subject more worthy of consideration any more than freckles on a human face. A consequence of such loftiness towards the problems of nature was a naïve amateurish philosophy of uncritical worshippers of science who, with all the fervor of prophets of the new epoch, were trying to replace the substantial arguments with wishful thinking. While philosophers locked in ivory towers were completely absorbed in quest of an objective spirit, the solution of many basic philosophical problems was left to engineers, as Comte, or amateurs, like Engels, or enthusiastic champions of the new epoch, such as Haeckel. The effect of their interpretation was a new vision of the world, in many ways comparable with the world of H.G. Wells' heroes.

Wells' short story *The Country of the Blind* illustrates perfectly how different may be interpretations of the same events. Far in the Andes, Wells' hero, Nunez, discovers a valley of blind people who do not even suspect the existence of the sense of sight. Contrary to the proverb that among the blind a one-eyed can be the king, Nunez is confronted with great psychic recalcitrance of his audience when he starts telling them about a reality of flowers, meadows, sun, and rainbow. His enchantment and enthusiasm are received as a humbug or evidence of his mental deficiency. Nunez is unable to prove his superiority through result obtained from the application of sight, since the blind have an excellent sense of hearing which enables them to solve all problems of everyday life. When he wants to marry one of the blind girls, he is told that this might be possible only after his consent to undergo a blinding operation. According to the blind medical experts, Nunez's mobile eyeballs were the cause of his mental disorder which led him to telling stories of an imaginary world of colors, light and flowers.

The marriage of Western civilization to modern science may promote the formation of attitudes similar to those of the Wells' blind. The criteria of truth that were imposed by pragmatism, favored, and still favor, focusing our attention on selected aspects of reality. Fragmentary, specialistic thinking facilitates the treatment of fields beyond the specialty as unimportant. From here, it is only a short way to considering certain types of reflection as expressions of poetical fiction, having no meaning and no counterpart in reality. Our daily practice and, connected with it, our style of thinking, influence our intellectual perception of the world to a much greater extent than we would be ready to admit ourselves. It does not seem to be a matter of coincidence that, for example, a decided majority of logicians working on set theory are Platonists.(1) Nor it is without reason that quite contrary philosophical sympathies dominate in circles of engineers and technicians. Like the blind in Wells' story, every day we are getting used to a certain way of interpretation. Sometimes we need extraordinary experiences, such as going through fascination, loneliness, death of a close person, to get out of the sphere of pragmatic hold and to notice another, new reality.

The fact that our vision of the world is an effect of phylogenetic conditioning, due to which the present shape of our species has been formed, calls for separate consideration. If we had at our disposal an organ of sight like a bee's eye or a magnetic receptivity of some birds, the horizon of our experience would be different. The poets would be enchanted by other impressions, and another would be the range of our ultimate questions. So far, the picture of the world formed on our retina depends on the fact that a human eye reacts to the length of light waves at the interval of 4000-8000 angstroms. Our sense of hearing reacts to sounds of the frequency between 16 - 20 000 Hz. Beyond the range of receptivity that we were alotted by evolution, there is an exciting mysterious world of whose existence we can only learn in an indirect way with the help of rational reflection.

One of the features of nature that arouses particular interest of theoretical physicists is its susceptibility to mathematical description. This feature seems even more intriguing when we realize that very often certain branches of mathematics were developed with absolutely no regard to their practical application. Apollonius of Perga, for instance, presented in 200 years B.C. his theoretical analysis concerning the cross-sections of a cone. This was, however, of no interest to astronomers of that time, as they were convinced that planets were revolving circles and epicycles. It was only 18 centuries later that Johannes Kepler discovered that orbits of planets are elliptic in shape and that their mathematical description can be found in Apollonius' works. Similar situation emerged in case of non-Euclidean geometries, matrix theory or group theory. At the desk, man created a new language, treating it in purely theoretical categories, and then it unexpectedly appeared that nature conducts a dialogue with us using this particular language. If we discovered an African tribe and found that its members recited fragments of Joyce's *Ulysses*, such fact could not be considered obvious and natural. Perhaps the people who do not know English or are always critical about Joyce, would not find anything strange about that situation; a sequence of English or English-like words would be for them only an unintelligible jabber. A similar situation occurs among humanists who express their astonishment that possibility of mathematical description of physical processes should be of major importance.

The mathematical character of nature by no means concerns such banal facts as adding together all consecutive sunsets or different colors of rainbow. It manifests itself in a non-chaotic character of certain physical processes. The universe might have been a sequence of uncoordinated events, and then it would have been impossible to distinguish in it the stable relations and universal laws that are easily described in simple mathematical formulae. If, for instance, in Newton's law of gravity the dominator had had a coefficient 2.001 instead of 2, our physics would have appeared much more complicated. If, in addition to the mass of bodies, when measuring the gravitational potential, we would have had to consider a progressive or reactionary character of societies in which the research was being done, the complication would have been still greater. It is not difficult to imagine a world in which the laws of physics would change together with political declarations or moods of overly sensitive people. So far, however, our world has been a world of relative stability and order, in spite of difficulties with mathematical description of our evolving emotions or of the motion of leaves blown by the autumn wind. This element of relative stability and order, rationality and harmony, appears both amazing and mysterious. It is because of its presence, a variety of physical processes can be described with identical mathematical formulae, no matter whether the processes take place in Leningrad, Pekin or New York.

The musing of scientists upon harmony and mathematical character of nature shares some elements with mystical experiences. Certainly, the experience of such harmony does not exclude the awareness of the existence of violent disharmony and of laws of jungle in nature. The disharmony may be, however, explained by pointing out to its immediate causes. The harmony leads us to a secret reality of rational structures which escape the attention of laymen but evoke passionate interest of specialists. When describing such a passion Albert Einstein wrote: "What a deep conviction of the rationality of the universe and what a yearning to understand, were it but a feeble reflection of the mind revealed in this world, Kepler and Newton must have had to enable them to spend years of solitary labor ... . It is cosmic religious feeling that gives a man such a strength. A contemporary has said, not unjustly, that in this materialistic age of ours the serious scientific workers are the only profoundly religious people."(2)

A scientist described by Einstein, like Nunez from Wells' short story, can discover a world of depth unperceivable for others. The contemporary drama of a split culture has its hidden source in the fact that comparatively few people can maintain the Einsteinian depth of perspective while investigating natural phenomena. In intellectual circles distant from the natural sciences, the standpoints of materialism and atheistic naturalism seem to attract the attention of scholars. Its presage may already be found in Samuel Clarke's polemics with Leibniz where two discrepant visions of nature crossed, each of them finding new advocates. An interpreter of Newton's texts, Samuel Clarke, considered the acceptance of God's existence indispensable to explain why catastrophic events which seemed a natural consequence of the laws of physics, do not occur in nature. Leibniz's God was not a cosmic sentry keeping guard against all sorts of disasters by suspending the laws of nature, but rather the God of pre-established harmony Who expresses his power in the regularity and order of cosmic processes.

Many representatives of Christian thought have yielded to the temptation of Clarke's simple explanations. His hypothesis of God was an easy way to remove all flaws and limitations of physical theories. When, however, the next period of investigation provided evidence that these flaws could also be evaded on natural grounds, without appealing to God, a question arose whether the hypothesis of God's action in nature was not merely an attempt to mask our ignorance. Various versions of this question appear in works of the authors who, like Leibniz, assumed that God's manifestation in nature should not be sought in gaps of our knowledge, but in the mysterious and fascinating harmony of nature.

The development of modern theoretical physics, the cosmological analyses concerning the so-called anthropic principle or the fluctuation of quantum vacuum, all show the harmony of the universe. The harmony described in the language of mathematical symmetries is much more expressive in modern handbooks of physics than in Leibniz's metaphysical comments on the pre-established harmony. This results in opinions that all discoveries in theoretical physics are in fact discoveries of hidden symmetries which in a different language were already described by Plato or the Pythagoreans.(3) In the context of similar opinions, the question arises whether the mathematically described order of nature cannot be treated as a manifestation of God's immanence in our world. Such approach has been particularly developed in the works of Teilhard de Chardin who, in his *Hymn of the Universe*, writes about the Divine as "that which in everything is above everything." This formula joins the traditional doctrine of God's immanence and transcendence. In another fragment of this hymn, the Franch Jesuit writes of God who is "so much remote in ... immensity and so much deeper in the intimacy of ... indwelling than all things else" and who unites "together the immensity of the world and the intimate depths of my being."

When approaching nature as a field of discovering God, who is the closest although infinite, Teilhard wrote in *Le Milieu Divin* about "divinity appearing at the heart of the universe." This philosophy raised as many objections as fascinations. When the widely known reviewers suggested to consider Teilhard's *Phenomenon of Man* a book of the century, Sir Peter Medawar replied in *The Mind* that the book was nothing but a collection of nonsenses evoking in its readers a feeling of disgust.(4) Many reasons could be given to explain such a great divergence of opinions. The presentation of optimistic hypotheses as basic truths, connecting poetry with scientific theories and overdone speculations, is not the least important. Such features could highly impede the apprehension of these elements of Teilhard's philosophy which seem to be most valuable and inspiring. These elements contain both mathematical description of cosmic order and the mystical experience of cosmic beauty. The synthesis of these two factors discloses a possibility of bridging the gaps that exist between the humanities and the natural science, between technological pragmatism and religious contemplation.

5

God Hidden in Nature

Teilhard's vision of God's presence in nature remains close to philosophical theories that are proposed in the so-called neoclassical theism. This influential school of metaphysics develops philosophical ideas originated by Alfred N. Whitehead, the co-author of the monumental *Principia Mathematica*. While Teilhard's readers might feel discouraged by his poetic metaphors, the readers of Whitehead must force their way through the barriers of his hermetic language, full of neologisms. This strange jargon provides, however, an exciting philosophical vision. The source of Whitehead's philosophical experience is the amazement of the fact that the universe is not a chain of chaotic processes devoid of regularity and sense. Before he took up creating his own metaphysical system, Whitehead had worked on the foundations of mathematics and principles of relativistic physics. Independently of Einstein, he presented his own version of the theory of relativity. While Einstein expressed his amazement that we could comprehend the universe at all, Whitehead tried to rationalize the amazement which results from the fact that the laws of physics do not change like freaks of neurotics or declarations of politicians. A universe in which the laws of nature would have a status similar to the so-called laws of social sciences would be easy to imagine. In more progressive societies, thanks to the enthusiastic unanimity of its citizens, the gravitational constant would assume a higher value; in societies showing little vigilance or zeal to fight for peace and progress, this constant would diminish by half of its normal. Such a vision of nature would undoubtedly satisfy ideological doctrinarians and provide them with physical confirmation of their "only just" ideas. So far, their interpretative diligence cannot lead, however, to socially dependent modifications of the universal laws of physics.

The laws of nature appear to be independent of social and cultural conditions, although one cannot determine reasons of their relative invariance. The very question why the laws of nature are invariant does not belong to questions that could be formulated within science. Science assumes the existence of certain constant relationships as a basic fact and does not try to explain them further. When expressing his amazement on the mysterious order of nature, John A. Wheeler wrote in 1974:"the miraculous identity of particles of the same type must be regarded as a central mystery of physics."(1) The identity of particles, the stability of relationships, the effectiveness of mathematics in description of physical processes yield many scientists to metaphysical comments. In his reflection upon these issues and in aesthetic meditation on the growth of wild flowers, Whitehead adopts the style practiced by William Blake in his poetry. In spite of important differences in their cultural and scientific background, these two authors in their reflection upon nature try to discover traces of infinity and to regard the harmony of cosmos as a manifestation of a hidden divine logos. The logos determines the direction of a potential development of nature and imposes certain restrictions on this development. Because of these restrictions, certain possibilities of the evolutionary growth are excluded from the domain of nature, and only in our dreams we imagine a different world with different laws of physics and peculiar ontology.

God's activity in nature does not mean that everything is determined by God. The acceptance of divine logos in the evolving world neither excludes the activity of creatures nor changes our world into a cosmic machine operated by a Clarkean mechanic. To the contrary, the process of the development of nature is the effect of cooperation between the finite, limited creatures and the infinite Creator. In Whitehead's metaphysics, God is not a transcendent immovable Mover, like in the metaphysics of Aristotle, but the Poet of the World who introduces His own patterns of harmony, order and beauty in cosmic evolution. The evolution cannot be, in the spirit of Teilhard's optimism, considered the necessary realization of these patterns, because it is not strictly deterministic in character. The lack of rigorous determinism is especially visible in the domain of human evolution. Owing to the ability of reflection and choice, we can affirm the ideals of development revealed by God or can also look for private self-made ideals, rejecting suggestions of the Divine Poet. A consequence of the latter decision will be illusory ideals that provide guiding principles for the attempts to construct new canons of private rationality and personal harmony.

In the framework adopted by Whitehead in his *Process and Reality*, the evolution of nature appears as a process of continuous creation. A subtle, invisible God manifests himself in harmony evoking our aesthetic enchantment as well as in the stable relations described by mathematical formalism of scientific theories. It is He who incorporates "our suffering and pain, our falls and triumphs ... in a continuously developing universal harmony." (2)

In the works of John B. Cobb, one of the followers of Whitehead, God appears as the ontic principle of creativity and progress in nature. God, Who in a special way inspires our activities, is the One Who calls us and wants to lead us beyond the immediately given reality toward what we may be in the future.(3) Such vision of God's cooperation with the created world guides us both to fascinating anthropology and to the affirmation of nature which, in a metaphoric language used by Charles Hartshorne, is described as "the body of God." In the attempts at theological interpretation of nature, a disjunctive alternative appears: either we assume the omnipresence of God in nature or, following Clarke's dubious philosophy, we will exclude a transcendent God from the domain of physical necessities and consider nature as a closed self-consistent system. Biblical premises clearly confirm the first indicated solution and disclose obvious simplifications in the traditional conception of the Transcendent Mover from the empyrean heaven.

The thesis of God's immanence in nature was developed also by many authors belonging to the tradition of Scholastic philosophy. Their arguments, however, had sunk into obscurity at the time when the positivists denied value of theology and claimed to be the supreme authority in the field of the natural sciences. Setting aside the cognitive pretensions of positivism, one can easily indicate the texts of Scholastic authors developing the doctrine of divine immanence in nature. A critical exposition of this doctrine may be found today also in the works of many authors who do not accept the tenets of Whiteheadian process metaphysics. Arthur R. Peacocke, a famous biochemist from Clare College, Cambridge, expounds in his *Creation and the World of Science* his version of the thesis of divine presence in nature when he argues:"God is in all the creative process of his creation ... everywhere and all the time present and active in them as their agent. They are his acts... All this is what we mean by saying that the Creator is immanent in his creation, and that is why we look for his 'meanings' (his intentions, proposals and purposes) within the world of which we are part."(4)

Peacocke's approach is currently enjoying particular popularity in circles of philosophers who examine problems of biological evolution. The acceptance of a rational principle of the dynamism of nature makes possible the evasion of many former controversies connected with the dilemma "either necessity or purpose", and enables rational explanations of the ultimate causes of biological evolution. Among Christian theologians, the doctrine of God's immanence in nature met with special approval of authors who investigate the problem of the role of the Holy Spirit in the created world. In John Taylor's theological reflections (5), the Spirit leading to complete truth and higher perfection resembles in many aspects the Whiteheadian Poet of the World.

An interdisciplinary approach, connecting the premises of the natural sciences, philosophy and theology, found its fullest exposition in the works of Charles Hartshorne. Thorough analysis of his works, inspired by theological content of the Gospel and the Epistles of St. John, originated the new theological program that is called neoclassical theism. The basic goal of this program is a search for the unification of religious faith, rational arguments and human emotions. Hartshorne's God, discovered in intellectual efforts as well as in aesthetic fascinations is both the God of logical inferences and of cosmic beauty.(6)

6

Rationality and the Drama of Golgotha

In contemporary attempts to develop neoclassical theism, a traditional split between the God of faith and te God of philosophers corresponds to a linguistic split between the language of Gospel and the language of theistic philosophy. With regard to traditional philosophical considerations on substance, invariability and perfection, the doctrine of the drama of crucifixion, revealing God's love to man, may seem both strange and bizarre. There are substantiated reasons to ask the question: If in the Palestine of Christ's days, the medieval patterns of classical theism had been accepted, would not the theological experts have questioned the doctrine of Incarnation and Holy Trinity, regarding it as discordant from the thesis of simplicity, invariability and perfection of God's being?

In the intellectual climate of ancient Greece, declarations about God's love for man sounded equally absurd as for us stories about square circles. In conceptual perspectives close to Aristotle, love was considered an expression of the creature's aspiration from less perfect to more perfect being. Consequently, in the sphere of the influence of ancient Greek philosophy, the concept of God as the immovable First Mover seemed natural, while St. John's presentation of God as the Person loving man and anxious about our problems appeared as intellectual dissonance.

The God of classical theism is much closer to Aristotle's texts than to Gospel. This is by no means an accusation, because in a general appraisal one has to acknowledge the important role of the Aristotelian-Thomistic tradition for philosophical expression of truth about God. Neoclassical theism does not, in any way, question the contribution of this tradition. It tries only, by enriching philosophical vocabulary and considering additional aspects of God's being, to express better the truth about the Infinite God. The quest for new theological patterns resulted in interpretations in which the world is presented as included in God. In this approach God, in a sense, shares the pain of his body and in his divine categories experiences the sorrows and sufferings that mark our human existence. (1)

The authors who speak about the co-suffering God stress the analogous character of the expression "divine suffering". Because of this analogy, we can neither put an equal sign between suffering of God and suffering of man nor regard God's suffering as a sign of ontic imperfection. The authors argue also that the doctrine of the so-called patripassianism, propagated at the end of the second century, has been rejected by the early Church not because of its ascription of suffering to God, but because of the false theory of Incarnation.

The conception of co-suffering God has been developed by François Varillon in his *La souffrance de Dieu*. He questions there the basic alternative of classical theism, according to which God is either imperfect and suffering or perfect and not suffering. Already at the level of human existence, we can find situations in which pain experienced in self-sacrificing love neither eliminates the feeling of happiness nor proves our imperfection. On the contrary, our susceptibility to other people's suffering, and the ability to share their pain are signs of our perfection. The ideal of personal perfection is not an insensible, impervious creature. Perfection manifests itself in sympathy, co-suffering and opening of a person to others.

In Hartshorne's presentation, the thesis of the co-suffering God finds a different justification than the one offered by Varillon. When we accept a theological model in which the world is treated as the "body of God", the suffering of the world becomes the suffering of God, for the same reason for which the pain of our body is our pain. Suffering of the world is a result of various mutual limitations characteristic of creatures which are marked with the stamp of imperfection. In this suffering, however, our imperfection is united to the subtle presence of the loving God. God, as the most perfect and most subtle being, experiences our failures in his specific categories. Strictly speaking, the failures are not exclusively ours, for because of God's immanence in the world, they are related to the being of hidden God.(2)

God's solidarity with mankind in our suffering is particularly revealed in the mystery of crucifixion. The drama of Golgotha remains a unique example of the unification of divine and human pain. Its union transforms the tragedy of sorrowful death into an expression of love and reveals the deepest sense of this love. The example of Golgotha is to remind us, in moments of our own suffering, of the co-present immanent God. His trancendence is the basis of our hope for the future crossing of the Rubicon that determines the boundaries of the domain of our pain. This step beyond the painful reality of the presence, the step towards the fullness of future existence, brings a trait of paradox in our life and introduces a hierarchy of values completely different from that adopted in naturalistic vistas. It is the hierarchy of order which is above our earthly existence. We discover manifestations of this order when we do not focus our reflection upon the surface experiences but, overcoming pain, we try to look for the depth of the hidden meaning of events.

Representatives of neoclassical theism do not try to justify suffering. They question, however, the ontic division between the insensible God and the suffering man. God revealed in Jesus of Nazareth chose the suffering of crucifixion as a means of redemption and a means of expressing truth about his love. The suffering of God in Christ, although shocking for philosophers, appeared to be a turning point in the history of mankind and provided an entirely new perspective for our existence. The world's suffering still remains shocking to our sense of rationality, both in terms of theoretical considerations and individual painful experiences. In spite of God's immanence, the world is still participating in pain, which is a clear sign that our quest for fullness has not yet been fulfilled. The human power to involve one's pain in the process of creative transformation of the world gives our existence a trait of moral beauty in which a suffering man creates human values in a much more difficult situation than a scientist or an artist. The heroism of creative suffering is evinced in the fact that we do not lose our faith in beauty even when we experience tragic disharmony between the horizon of theoretical possibilities and actually attainable actions.

The concept of moral beauty may seem to be a symptom of unsuitable aesthetizing while facing pain. Before, however, raising an objection of this kind, one should consider the axiological-ethical consequences of such an approach. It cannot be denied that our civilization highly values the harmony of logical inferences expressed, for instance, in comprehensive scientific theories as well as the harmony of beauty present in works of El Greco, Mozart or Shakespeare. Much more difficult than the harmony of these works and, at the same time, underestimated in our civilization, is the harmony of human existence in which the experience of pain does not destroy the experience of meaning. Such attitude is much more difficult as, while in the creativity of an artist or a scientist, essential is only a one-aspect harmony of formal constructions, in suffering a much more complex harmony is needed which embraces all aspects of human existence experienced by the suffering person. If in the cultural development of mankind, the Socratic elements had been of greater importance, we would, perhaps, have been able to perceive the symptoms of genius and greatness in a lonely struggle with suffering. Our civilization, however, has inherited different criteria of valuation which are frequently in flagrant disharmony with the thesis of a rational character of human nature.

All attempts at a general-theoretical explanation of the phenomenon of suffering may appear abstract and hollow when confronted with individual human pain. There may always be objections that omnipotent God could have stopped the development of illness or prevented a tragic accident. In works written by the followers of Whitehead's and Hartshorne's ideas, the attention is concentrated not on hypotheses of the most perfect of possible worlds but on the question how to create the global harmony of sense in our existence marked locally by pain and suffering. In such vistas, the system of reference connecting God's immanence and transcendence, transforms our tragedies into constituents of a higher, more subtle harmony of human existence. The split introduced by suffering into our existence acquires a new sense within this very harmony.

7

God Omnipresent Incognito

It has always been a dream of mystics and masters of spiritual life to view the whole reality in relation to God's being and to perceive God's presence in the context of ordinary prosaic events. A similar goal was the guiding principle for contemporary philosophers, searching for means to prevent splits between science and faith, the humanities and the natural sciences, the God of philosophers and the God of Israel. A philosophical standpoint accepting God's presence in nature and, at the same time, God's transcendence towards the world, is known as *panentheism*. Negative consequences of using this term are doubts dealing with consistency of the doctrine of *panentheism* with religious orthodoxy. The doubts arise because of psychological association of *panentheism* with pantheism. According to J.A.T. Robinson, many sympathizers of *panentheism* hesitate to declare openly for this doctrine, fearing possible accusations of heterodox pantheism.(1) When such accusations appear, it is a result of sheer misunderstanding, because *panentheism* is the fullest, consistently developed form of classical theism.(2) When it is not associated with additional hypotheses which would, for instance, question the contingent character of the created world, *panentheism* remains in perfect agreement with the assumptions of natural theology that were approved in classical theism.(3) Its implicit anticipations may be found in classical religious texts where elements of the biblical vision of nature complementarily intermingle with verses of poetry describing "a subtle sense penetrating into depth" or "a source of freshness hidden in the depth of things."

The perspective offered by *panentheism* is not only intellectually inspiring and aesthetically attractive; its important trait is also the close relationship with the biblical perspective. Contemporary sympathizers of *panentheism* point out that already in Judaistic pre-Christian philosophy one could find numerous examples of reflections about God "penetrating" nature and revealing his presence in prosaic events; not only in extraordinary miracles but also in ordinary situations of everyday life. Ruach Elohim, the spirit of God, is present in the created world since the very beginning. In the biblical description of creation, he introduces order, determines cosmic laws, controls the process of the development. God penetrating nature and the horizon of our daily experience was the God of the author of Psalm 139, who wrote: "O Lord, you have probed me and you know me; you know when I sit and when I stand; you understand my thoughts from afar. My journeys and my rest you scrutinize, with all my ways you are familiar." (Psalm 139:1-3).

St. Paul had a similar image of God when, in a culturally alien environment, he spoke to the Greeks of God in whom "we live, we move and have our being" (Acts 17:28). Such statements are testimony to the exceptional maturity of Paul's natural theology. God about whom he spoke at the Areopagus was no longer a God of one of wandering tribe, but the God of the universe in whose existence is rooted the being of each of us.

The term "to be" is one of the most misleading verbs. We "are" in a different way through our works and our thoughts, in a different way in a photograph or in one's memory, and in still another way through physical presence that is combined with spiritual distance. Beside these various types of presence, there is also the fullest presence combining psychic union with physical closeness. God's presence in the world may also assume various forms. Its fullest manifestation are the sacraments, another is prayer, an experience of moral goodness or aesthetic fascination. In the consciousness of the early Church, not only the truth of God present in the community of prayers was alive, but also the truth of God penetrating the existence of individual creatures and the domain of natural phenomena.

The doctrine of the Holy Spirit reminds us of the truth about God who is the God of immanence, who permeates, purifies and strengthens the human interior life, bringing transformation into the scale of man's valuations and moral principles. The Spirit integrating the congregation and supporting its activities is the God of harmony, whose presence may be discovered in the whole cosmos. The water of Jordan was not only meant to fulfill pragmatic functions but was also a symbol of God's purifying power. Bread and wine, needed for the fullest presence of God in the Eucharist, provide the most significant confirmation of the theological significance of nature. It is a matter of cardinal importance that they are not simple fruits of the earth, grapes and wheat, but their forms processed by man who "fills the earth and subdues it" in cooperation with God "transforming the face of the earth."

The elements presented above show that the Christian attitude towards immanent God cannot be reduced to passive contemplation or simple discovery of God's vestiges. A Christian's task is the cooperation with God revealed in nature and in the depth of our personal feelings. This cooperation cannot be, however, properly realized when God is only regarded as an abstract, transcendent being. Such approach could be considered an implicit form of a return to deism. The indispensable condition of the proper development of cooperation between human and divine elements is the connection of human our freedom and of ideals of harmony that are proposed by God as an ideal and model of our existence.

Respecting our freedom, God does not act through strict determination or an imposed constraint, but through persuasion which presents possibilities of full realization of our humanity. Man is free to reject God's patterns of harmony and to construct private ideals which assume private vision of rationality. So, the chosen hierarchy of values may be either an expression of our cooperation with immanent God, or an effect of the rejection of his ideals and directing our life towards different values. In both cases, the immanent God still remains in the human being, but there are two essentially different forms of his presence. The discovery of this presence may proceed in different ways, depending on a paricular existential situation of the searching person.

The latter of the situations mentioned above shows how the elements of ethos and rational search are joined and mutually conditioned in our existence. Like an observer in quantum mechanics, who out of stern necessity is a participator connected with the observed phenomenon, so do we, in our quest for God, form a system in which the elements of logos and ethos are united in one, mutually dependent harmony. This symptomatic union finds a special expression in lives of the people who have not yet formally accepted the thesis of God's existence, in spite of the fact that their lives are already a practical realization of ideals and patterns revealed by God. In such cases, even if a searching person still regards himself as an atheist, there are reasons to regard him as an *anima naturaliter Christiana*, for his way of living is the result of a, yet unconscious, cooperation with immanent God in the realization of His ideals determining the harmony of human life. In such union, God in whom "we live, we move and have our being" penetrates us, knows us and guides us toward himself, even then when we are still hiding in the well-ordered world of our personal stereotypes.

Panentheism provides an important form of opposition against philosophical simplifications expressed in what I call misplaced divinity. Traditional theism, because of its stressing the truth of transcendent God, accepted as necessary evil the split between the Abstract Absolute and human personal anxieties expressed in different ways in Abraham's life, Augustine's reflections and Kierkegaard's dilemmas. In such approach, God was eliminated from nature and from the field of mundane experience. A different form of eliminating divinity was provided by atheistic naturalism, in which nature was treated as the domain of naturalized divinity without transcendence. In the latter perspective, reflections on the beauty of physical theories or poetry of sunsets could only result in Camus' tragic humanism or in Sartre's existential absurdism.

Panentheistic synthesis suggested by neoclassical theism provides a form of counteraction against the flaws of the two mentioned solutions. From its point of view, crucial is the truth of God present incognito. The area of experiencing pain and the field of aesthetic enchantment determine the horizon of God's subtle presence. Abstract formulae of physical theories remain in direct reference to "the body of God," which the universe is. In this way they acquire both anthropological and theological character. The world of our ideals and human values becomes the area of mutual interpenetration of divine and human elements.

The discovery of God present incognito is accomplished through rational reflection on nature as well as through contemplation and meditating upon ethical-axiological aspects of our existence. Rooted in the world and sensitive to its beauty, we discover a hidden dimension of our being through a cognitive efforts which reveal the element of logos, hidden in the course of passing events. This attitude of sensitive openness to rationality, poetry and mystery of the world leads to a new reality of "spiritual experience". Its manifestation is the discovery of a deeper character of things and our union with them. Such experience may be described in psychological categories of sublimity, internal liberation, awe, peace, etc. Feeling bonds with other people, experiencing inward peace, changing valuations and values, discovering new possibilities to realize harmony in our life constitute the field of our immediate approach to the divine reality of the immanent God.

Speaking of God's Kingdom, Christ did not compare it to cedars, captivitating with their impressiveness, or roses, fascinating with their beauty. Inconspicuous charlock and unattractive but effective leaven were used in metaphorical comparisons to reveal the essence of God's Kingdom, the essence which is expressed through the small, prosaic and unnoticed. The mystery of the action of divine grace reveals its beauty to those who, like the little Prince of Saint Exupery, remember that the most important is invisible.

The concept of God proposed in *panentheism* results in important theological consequences. One of these is the treatment of human being as a fragment of God's life. Our existence may be described in conceptual categories of different scientific disciplines, but its full characteristic, free of splits and contradictions, is accomplished only after the consideration of theological aspects of our being. Certainly, one can exclude these aspects already at the starting point by a dogmatic rejection of natural theology, the rejection based on dogmatic assumption that only the natural sciences are the source of valuable knowledge. In a similar way, one could dogmatically declare that our relatives are, for example, only the sets of vibrating electrons, and all attempts to describe them in terms of emotional links, aesthetic or ethical categories are barely an expression of naïve anthropomorphisms.

The choice of the last of the mentioned attitudes would mean the mutilation of the world's richness by the amputation of its most relevant constituents. Neoclassical theism, particularly in its *panentheistic* version, is an ambitious attempt at uniting separated elements, so that they formed a full and free of contradictions vision of our world. Although some of the detailed suggestions still require fundamental revisions and further development, such approach appears as an interesting proposal of counteracting the splits of our civilization through the great unification undertaken in natural theology.

8

Waiting for a New Ferguson?

The split between the scientific and the religious vision of reality is to a great extent the result of the 19th century's intellectual climate. Still in the 19th century, a Scottish astronomer James Ferguson at his public lectures argued that one should harmoniously connect the results of astronomical discoveries with the religious truth and the existential reflection on the beauty of the world. His synthesis was received with great interest outside scholarly centers when Ferguson informed delighted listeners that some nymphs prefer astronomy than love since the former raises human spirit into the highest regions of the universe. If Ferguson lectured today, he would have a more difficult task to fulfill, not only because the arguments refering to nymphs have lost their strength. The current development of physics requires with inevitable necessity the elimination of external interpolations of moral or theological character. One should thus not wait for scientific Fergusons, because this waiting would be a waiting for Godot.

Certain contemporary authors, indeed, dream about the possibility of deep transformations and revisions in the present scientific paradigm. The revisions are to introduce some existential problems into the domain of scientific investigations. Up till now, however, all atempts of a priori forcing upon science normative principles of its future evolution appeared failures. There are serious reasons to argue that the expectation of a future unity of the natural sciences and the humanities remains only a nice dream which has no chance of fulfillment. Instead of comforting oneself with the expectation of the internal evolution of science, it seems to be better to develop philosophical reflection upon factors that divide the scientific thought from the religious one. Such factors are certainly mutual misjudgements and prejudices. The gap between scientific and religious vision of the world is to a great extent the result of the fact that in clasically understood natural theology the thesis of God's immanence in nature did not obtain sufficient attention. The efficient counteracting against the current atomization of cultures requires both the development of this thesis and pointing out that certain prejudices are unsubstantiated.

Among various objections put forward against Christian vision of nature one quite often can come across an opinion that both the domination of technology on humanistic values that occurs in the West and the terrifying ecological devastations have their intellectual source in an instrumental subordination of nature to man, the subordination essential for the Judeo-Christian tradition. The supporters of such an approach relate to the biblical command:"be fertile and multiply; fill the earth and subdue it." (Gen 1:28). They cite this text to oppose the Judeo-Christian mastering of earth with the attitude of contemplating nature, an attitude characteristic for Eastern thought.

There are serious reasons to doubt if managers who devastate carelessly the natural environment are really looking for inspiration in meditating on the Book of Genesis. The counterargument claiming that, in spite of the lack of direct influence, the biblical command shaped the general atmosphere of unhampered exploration of nature remains at least unverifiable. There exists no possibility of empirical testing to what extent the so-called general intellectual atmosphere is shaped by only one sentence of the Old Testament, taken in complete isolation of other sentences, and to what extent this atmosphere is created by the economic principles of new barbarians. In such a situation, the easiest pseudo-solution is to assume that God or the Bible are guilty for everything. If there were not the command in Genesis 1:28, the grains around Chernobyl would have flourished and children would have joyously played around the local power station.

The thesis of specifically western phenomenon of ecological devastation seems at least questionable because one can provide many examples of similar devastation in cultures where no religious premises were accepted to justify the instrumental treating of nature. René Dubois in his book *A God Within* presents a large collection of examples which weaken the position of simple explanations of present ecological crisis.(1) The examples of the Australian aboriginals' burning steppes, the Asyrians' destruction of large animals, the Romans' deforestrations are difficult to be reconciled to an idyllic vision of cultures that are uncontaminated by western technology and the Bible-dependent thesis justifying human domination on nature. The texts of Chinese poets glorifying the union between man and nature have also small power of argumentation since the authors of these texts were often landlords who could praise the cosmos at the same time when their workers conquered the earth not for themselves but rather for poetically-minded landlords.

It seems very risky to look for theological presuppositions of phenomena which can be explained in terms of barbarian economics. Without undertaking such a risk, I would nevertheless accept the thesis of the essential difference between the theological vision of nature accepted in certain religions of the East and the conception of nature that prevailed in Christian thought during the last two centuries. The evolution of scientific ideas, accompanied with the enormous technological development, led to the situation in which Christian philosophers practically acknowledged the domain of nature as an exclusive subject of the scientists and technicians. Such an approach stimulated the displacement of the attention of theologians and philosophers beyond the world of nature. Despite the fact that the truth of God's immanence was never questioned in theological reflection of our times, one talks more often about transcendent God than of God hidden in nature.

The contemplative insight in which one discovers the sacrum present in nature is made difficult not only due to a small number of the practitioners whose approach to nature becomes the terrain of discovering divinity in the world. This approach may be found in Simone Weil's comments about God who smiles on us through the material world as well as in mystical cosmology of Teilhard de Chardin. The theological basis of this approach was explored by Thomas Merton when in his reflections upon contemporary mysticism he postulated to "become detached from ourselves in order to see and use all things in and for God."(2) This kind of religious plunging in immanent God can be easily reduced to vague clichés about world's affirmation, as in the past it had been reduced to empty slogans about world's rejection. Avoiding the two easy extremes, Merton reminds us of the basic tenets of religious insight when he asserts: "it has never been Christian to reject 'the world' in the sense of the cosmos created by God, dwelt by the Incarnate Word, sanctified by the presence and action of the Mystical Christ, and destined to be transformed with man in a new eschatological creation."(3)

The problem is that the Mertonian approach seems incomprehensible for the admirers of the fascinating world of perfected technology, who have no concern for thelogical abstractions, as well as for self-complacent repesentatives of religious thought, who demonstrate but lofty disregard for the so-called terrestrial values. Members of these two groups, satisfied with themselves and with their inviolable principles, affirm in everyday practice a radical opposition between the truth of science and of the Gospel. In spite of the fact that, at the beginning of the 17th century, Galileo argued that God reveals himself both on the pages of the Bible and in the book of nature, now more than three centuries after him one can still find the radicals who suggest that a real intellectual aristocracy necessitates ignoring one of the mentioned books.

The alleged opposition between the revealed truth of the Bible and the truth of rational reflections upon nature found, among the Christian authors, particularly strong expression in the works of Karl Barth. A respect for his original theological contributions was expressed by inviting him to Scotland to give prestigious Gifford Lectures. The wish of Lord Gifford was that these lectures would contain analysis belonging to the domain of natural theology. In first words of his lectures Barth shocked the gathered listeners, when he stated laconically:"such a science as Lord Gifford had in mind does not exist."(4)

The radical Barthian approach does not come from the questioning of the thesis that God may reveal his presence in nature. It is rather the result of distrust to human intellect which can easily become an idol in civilization impressed by rationalism and belief in infinite cognitive capacity of human being. In order to escape away from the threat of modern intellectual idolatry, Barth declared the rule *solus Christus*, -- sole Christ, and argued that the only form of discovering the truth of God is the word of God revealed in Christ. Such a radical setting implies calling into question the rational legacy of the earlier Christianity. It is easy to point at the noble intentions and sublime motives of Barth's rejection of natural theology. Good intention, nevertheless, cannot replace substantive arguments. Connecting of the thesis *solus Christus* with the radical critique of human reason introduces deep inconsistency into Barth's theological arguments. In order to recognize the full sense of biblical teaching and to define the real meaning of the principle "sole Christ," we need both the rational reflection and critical analysis which distinguishes between the essential content and its accidental forms. Actually, Barthian program would not lead to a better understanding of religious truth but its amputation resulting from the lack of confidence in human intellect. A consequence of carrying out this program would be impeding the discovery of the sacrum hidden in nature and disregarding theological aspects of nature's beauty.

The position of Barth appears attractive for the persons programmatically critical towards rational reflections on nature. When one takes into account, however, how highly these reflections contributed to the transformation of our picture of the world, one can hardly find any objective reasons for raising the personal distrust towards natural theology to the rank of basic criterion of truth.(5) The need for theological reflections upon nature seems obvious. The problem remains, nevertheless, how to avoid in such reflections replacing substantiated arguments by poetical metaphors that stem from wishful thinking.

As late as in the 1950s some philosophers of science have pointed out that it is necessary to create a new discipline which would investigate the humanistic aspects of the natural sciences and simultaneously would be a bridge between domains which are separated with the obvious loss for our culture. In this bridging of narrow fields of specializations, the important role can be played by the theological reflection on the essence of science. Its results would belong to the theology of science. This new discipline should not be regarded as a product of new fashion for creating new types of theologies, unknown in the past. If we want to counteract effectively the intellectual byproducts of scientific specialization, we have to see an integral picture of various aspects of scientific research. Thus it is necessary to develop a theological reflection concerning science as a whole. Modern science is, for many reasons, an exceptional achievement of the human intellect. The fascinating feature of our species is manifest in the fact that we are able to discuss cosmological models, genes and quarks with the same passion that animated lower animals in their fight about food and a habitat.

The scientific reflection and the idea of truth connected with it have important anthropological and theological presuppositions. Those who concentrate their attention on technological aspects of science, flattening and mangling the richness of its content, are not able to notice theological dimensions of scientific research. Their surface image of scientific discovery facilitates radical changes of opinions concerning the essence and value of scientific research. As a result, after a period of uncritical enchantment by science there comes another period in which one is inclined to see in science a source of both moral evil and social failures. In this context, the theological assessment of science, if it avoids both the uncritical fascinations of the past and the simplifying disapproval of today, can play a very important role, provided that the assessment would be founded on religious conception of nature, the creation, human being and our cognitive capacity.

The basic premises for theological assessment of the natural sciences may be derived from biblical text which presents nature as a domain of discovering a deep hidden sense. In the biblical approach, the lilies and birds became the elements of the visible reality which points to the invisible world of grace. Vines and fig trees, the nets filled with fish and the experience of the storm on the sea, all this had a significant theological component, thanks to which God's reality entered into the usual human existence, revealing its new fascinating dimensions. Those who did not want to see it could have confined themselves to the simple registration of natural events. Those who accepted the religious perspective discovered supernatural reality permeating the beauty of nature. They followed the author of Psalm 104 who wrote about God present in nature:"O Lord, my God, you are great indeed! You are clothed with majesty and glory, robed in light as with a cloak. ... You make the clouds your chariot; you travel on the wings of the wind. ... You sent forth springs into the watercourses that wind among the mountains. And give drink to every beast of the field." (Psalm 104:1, 3, 10f). In such a perspective, cedars of Lebanon, plain charlock, high peaks of the mountains, goats and boulders, useful herbs and wine cheering up the hearts were part of the act of creation in which one discovers the presence of the divine Spirit who gives the new life to the earth (Psalm 104:30).

The presented perspective implies the conviction that the earth and its inhabitants constitute the field of discovering traces of God's presence. Immanent God close to us and searching our feelings was presented in the theology of St. Paul as the God of great cosmic unification. Christ of the epistle to Colossians appears as the image of the unseen God who connects what is in heaven and on earth, seen and unseen. In him everything in heaven and on earth was created, things visible and invisible, ... all were created through him, and for him. (Col. 1:16).

The same idea of the highest unification between the world of nature and invisible God's reality was developed in Augustine's philosophy. The field of human inner experience, personal searching and intimate reflections constitute in this philosophy the area of discovering the most fundamental truths. In his theory of illumination, Augustine uses the metaphor of light to express the truth that our anxiety and existential dilemmas constitute the domain of encounter with God who comes to take us from the play of attractive shadows to the true reality of his light. Seven centuries later the same philosophical undertone appears in the works of Anselm of Canterbury, who was called by his contemporaries the second Augustine. On the pages of the *Proslogion* Anselm expressed his feeling of the close bond with the immanent compassionate God when he wrote: how art thou compassionate, and, at the same time, passionless? For if thou art passionless, thou do not feel sympathy; and if you do not feel sympathy, thy heart is not wretched from sympathy for the wretched. ... How, then, are thou compassionate, o Lord?

In the legacy of later Scholastics, the element of rational reflection definitely dominates over Augustine's yearnings and emotional questions of Anselm. Nevertheless, in the rational discourse of Thomas Aquinas, the arguments for God's existence are founded on cosmological premises stemming from observation of natural phenomena. The world of nature, through its mixing contingency and harmony, enables the elaboration of the *quinquae viae*, the Thomist arguments for the existence of God. Human reflection on nature constitutes in this philosophical framework the beginning of the way to the discovery of God. God, understood in Aristotelian categories, is regarded above all as a Transcendent Cause or a distant Immovable Mover. At the same time, however, in mystical tradition continued by St. Bonaventure, God still has been presented as close to us and discovered in the depth of human feelings. Like in Augustine's works, He can be found in the union of prayer that discloses an invisible fundamental reality. The rational creatures are in a special sense God's images, because the immanent divine illumination enlightens and transforms human souls. "Christ" - argues Bonaventure - "is the interior teacher. No truth is known except through him, ... by his enlightening us interiorly ... . He is intimately present to every soul and by his most clear ideas he shines upon the dark ideas of our minds."(6)

Marks of the departure from the reflection unifying human feelings, rational arguments and theological content can be noticed in the epoch of the Renaissance. In spite of the fact that the conception of universal science, *scientia universalis*, was propagated in this period, the broadly understood science remained closer to the poetry of nature and to humanistic rhetoric than to the systematic study of natural phenomena. The essence of Galileo's approach remained incomprehensible for many Renaissance humanists who were convinced that contemplative reflection on nature was merely to satisfy a pious curiosity but not to deliver reliable knowledge. In this framework, human scrutiny of nature could have provided nothing but illustrative exemplifications to the biblical teaching. The poetic phrases of St. Francis' *Hymn to the Sun* remained close to this tradition; the tenets of new astronomy and physics remained beyond its reach. The heliocentric model of the world must have waited a long time for Fergusons who in its structure perceived disclosure of both human values and the divine Wisdom.

9

The Closed Society and Its Religious Enemies

> The idols against which the Old Testament prophets fought were idols in stone or wood, or trees or hills; the idols of our day are leaders, institutions, ... production... .
>
> Erich Fromm, *The Revolution of Hope*

Openness to Transcendence

When religious spirit is erroneously identified with religious formalism, one tries to oppose the essence of religious experience to the so-called "secular" values that are acknowledged in the natural sciences and the humanities. This approach results in the regrettable phenomenon of a religious ghetto. According to the inhabitants of the ghetto, closed religious communities are to be saved from vicious external influences. Their closedness is to stem from the very essence of religion. In certain periods, the mentality of the religious ghetto played a positive role that prevented some religious communities from the harmful influence of unsubstantiated speculations. This relative utility does not, however, justify the introduction of general evaluations. Criteria of pragmatics, important for economy, can hardly be decisive in religious assessments. The essential element of the religious attitude may be sought in human openness to a transcendent divine reality. The reality may appear in various forms; its weaker or stronger manifestations depend on particular systems of the religious beliefs that are adopted. The very core of religious attitude contains, nevertheless, the element of openness to a richer reality which transcends the domain of nature. This openness generates a specifically religious sphere of a radiation of holiness.

Regardless of terminological exuberance characteristic both of theological and philosophical discourse about God, one may regard the openness to transcendent holiness as a common element essential for various types of religious experience. In our psychic perception, this element often appears as coupled with the experience of fascination or fear, tenderness and joy, love, enchantment, or desire for fulfillment. In spite of the richness of their nuances, in all types of these experiences we discover the essential component which destroys our egotistical self-centeredness and directs our attention toward the domain of values that transcends the limits of our well-established personal world. In religious experience, as in the Popperian theory of society, the openness remains the decisive factor and a necessary condition of a rational attitude towards the world.

In influential religious systems, whether we consider Judaism, Christianity or Hinduism, the relation between human existence and the transcendent divine reality cannot be reduced to a private I - Thou communication. The divine Thou, when penetrating our inner self, radiates in the domain of our cultural and social values. God's transcendence underlying our experience of holiness transmutes the sphere of our intimate immanence. The transcendence that comes within the horizon of our personal privacy enlarges the domain of the immanent and discloses new aspects of our everyday experience. The well-established horizon of our self-centered world expands because of religious creativity which implies a necessary element of the dynamic growth.

Creativity inspired by religious reflection can hardly be confined to a closed universe of well-known experiences. The depth dimension characteristic of religious insight can neither be reconciled with axiological narcissism nor be described in a finite closed set of simple formulae. The richness of religious experience cannot be adequately expressed by means of accessible finite linguistic forms. When we introduce additional new forms, new relationships between religious and human factors are disclosed. Simple schemes reveal their deep content and give rise to important amendments in the set of assertions that were earlier accepted as ultimate truths. This regularity adopts in mystical experience different forms than in the rational discourse of natural theology. Nonetheless, neither the emotional bliss of mystics nor the logical deductions of philosophers can grasp the fullness of the divine.

The transforming creativity which is contained in our religious experience seems to be particularly important for the contemporary technological society. By focusing attention upon human tenderness, sensitivity and compassion, it opposes depersonalizing trends that exist in contemporary society and reminds us of the multidimensional reality of human existence that cannot be reduced to its pragmatic-technical component. The meditative practices of the East and the longlasting tradition of Jewish-Christian mysticism evoke awareness of these elements of the human inward life to which no attention was paid at the period of scientific enchantment. The experiences of human loneliness, love, or suffering cannot, for methodological reasons, be thoroughly explored on the level of scientific investigations. Their role, so important for human life, is recognized in religious perspective where one regards them as parts constituting a harmonious whole. The search for wholeness, promoted by fundamental religious questions concerning the meaning of life and the end of terrestrial existence, facilitates overcoming the patterns of behavior that prevail in consumer society. Human inner integrity and outward harmony with God and other people cannot be based upon the growth of GNP factors or upon unrestricted collecting of material goods. Priority of "to be" over "to have," before it became acknowledged in the philosophical analyses of Gabriel Marcel and Erich Fromm, was consistently proclaimed in various religious traditions where the spiritual emptiness of purely consumer existence was pointed out.

The awareness of the presence of mystery in human existence prevents both uncritical Promethean pride and technological idolatry in which science is to provide secularized redemption for the enlightened. Jesus' principle "man does not live by bread alone" expresses the essence of religious directing towards the transcendent reality. It makes us aware that our axiological horizon cannot be narrowed to a finite set of values even if they are important or necessary. In these cultures which focus their attention not so much upon bread as upon technology, pleasure or consumption, the principle mentioned above shields people from easy fascinations in which relative values are regarded as absolute or, at least, as the highest ones. In such cases, the religious openness to transcendence permits us to consider a larger historical perspective and prevents intellectual parochialism that is focused upon the immediately given. It is easier to overcome the temptation of such parochialism when we take into account religious premises that reveal the cosmic dimension of our existence and direct our attention toward the invisible transworld reality.

This redirecting of intellectual concern does not result in an alienating isolation, because the discovered divinity transmutes the horizon of our habitual experience. Its immediate cultural impact may be found both in internalized ideals and in external rituals. Myths and hymns, music and art, corporate prayers and personal experience of holiness reflect the religiously conceived transcendence that enters into the sphere of secular experience to enrich it and to define its new meaning. Our daily bread and the unnoticeable growth of God's Kingdom unite in harmonious conjunction and unveil boundless terrains for our personal development.

In religions of the East, where the sphere of the divine is not perceived in personal categories, similar integrative effects appear in the personal experience of harmony and peace that presupposes the underlying unity of all things. Such a unity, experienced in meditative insight, engenders feelings of inward fulfillment and self-realization. Such processes, whether they occur in Buddhist or in Taoist tradition, introduce the element of holiness into the world of natural order. This opens the world toward new ideals of sublime harmony. It is true that not always and not in all cultures can we indicate observable effects of such an opening. The effectiveness of religious ideals and their practical impact on social behavior have been, however, a separate problem.

On the level of theoretical analysis, I share the opinion of Erich Fromm who argues that dramatic changes would take place in our whole culture if the people would really follow the principles of the Buddhist Eight-fold Path or the Ten Commandments.(1) The painful split between sublime theory and annoying practice manifests more a fundamental split latent in human nature. It was Socrates who dreamed of the harmonious existence dependent on synchronizing the logical and ethical components of our nature. Twenty four centuries after his death, the great Socratic vision still holds the attention of authors who attempt to construct a consistent system of ethics based on rational premises that express the most valuable elements of the cultural heritage of mankind. Disharmonies noticeable in contemporary society protect us from thinking that an instant harmony can be easily achieved in any domain of human existence.

Spiritual Adventure versus Religious Formalism

The richness of the human experience of divinity cannot be exhausted by means of finite conceptual tools characteristic of particular religious systems. In our perception of the divine reality there are no theological counterparts of Mount Everest, since for any form of peak religious experience there always exist previously unknown routes directed towards exotic highlands of fresh spiritual discernment. Of its very nature, religious insight offers an element of intellectual adventure, an element that psychologically is remote for the mentality of the bureaucrats who enjoy attempts at the unsophisticated cataloging of the splendid world of human experience.

Relatively often, misshapen religious experience has been regarded as a domain of theological bureaucracy. Rigid formulae, in such an approach, seem to be more important than the splendid beauty of religious values; and sharp theological distinctions as well as colorless trivialities conceal the natural charm of the radiation of holiness. Similar effects occur as byproducts of attempts that aim to incorporate our experience of the sacred into systems of institutionalized religions. If the role of formal structures is overestimated in the systems, their depth content can be hardly recognized and adequately assessed. The same mechanisms function in many other fields of human axiological experience. Individual disappointment with the social structures of institutional science very often leads to total denial of scientific values. Personal frustrations caused by failure in the performance of given musical compositions can result in psychic resentment against music as such. When, however, necessary distinctions are adopted in all these domains, one must differentiate between an objective value in itself and its individual concrete actualizations. The flaws in the latter entitle no one to call into question the former.

Relics of the unrealistic sentimentalism of Rousseau often appear in contemporary yearnings towards "pure" and "natural" values that are free of the contaminating influence of modern civilization. The poetic component in such yearnings definitively dominates their rational justification. The very notion of nature, basic for today's versions of romantic neo-naturalism remains vague and facilitates profound misunderstanding among enthusiastic supporters of the attitude which is directed against institutions, industry and present cultural patterns. In certain versions of this philosophy we find nothing but dreams of the lost paradise of human childhood, the paradise which has no chance of being restored again in our industrial society.

References to human nature appear in philosophical comments that are mutually as distant as Thomist doctrine of the natural law and the sociobiological elaborations inspired by Edward O. Wilson's *On Human Nature*. The remarkable differences of opinions in understanding the key notion of nature are evident when the sociobiological explanations of human nature are presented as a continuation of philosophical ideas of authors as different as Hume and Plato, or Kant and Hegel.(2) Numerous skeptics who reject Wilsonian tenets argue that the sociobiological theory of nature suggests the jungle to be the most natural milieu for the well-mannered intellectuals from Harvard. Setting aside similar disputes, I would like to stress merely that our yearnings for so-called natural values, the values exempt from the destructive influence of technological culture, remain attractive until we try to define the basic notion of nature. This notion appears so ambiguous that its users are free to ascribe an essentially different meaning to the same naturalistic phraseology.

At the end of the 19th century, in the intellectual climate generated by scientism and positivism, many uncritical slogans were repeated to praise technology and science for their dubious merits. Technology was extolled at that time because of its apparent achievements in preventing mankind from the tyranny of nature. Today the fashion has changed and nature is extolled as a means of liberating the human species from the tyranny of technology. In both cases, the very notion of nature seems equally fuzzy and it needs fundamental logical elaborations. Logical distinctions and the elementary requirement of defining the basic consepts destroy the provocative charm of the new naturalism. Novel versions of Rousseau's romantic philosophy seem to be as appealing and ungrounded as their 18th century archetype.

The desire to contemplate "pure" values free of social-cultural contaminations seems then both attractive and surreal. Axiological realism should be manifested not in utopian nostalgia for the paradise lost but in looking for the effective means that would enable axiological reorientation of our species toward values facilitating human fulfillment. The very concept of human fulfillment has obvious philosophical undertones. Concrete conceptions of fulfillment depend on our general Weltanschauung and imply theological as well as ontological commitments. In the variety of theoretically possible solutions, religious theories of values provide important premises to develop the humanistic tradition of ideals, the tradition founded on deeper spiritual values. Our capacity to transcend the narrow domain of human loneliness, cosmic despair, prisons of selfishness, can be strengthened when we are attentive to the religious vision of the open society. Its basic openness stems from the fact that the transcendent divine reality may attract human efforts manifested in our quest for a meaningful life and for a consistent hierarchy of values.

Various religious systems provide different answers to the particular questions concerning the relationship between religious and the so-called secular values. Even the basic question of the relationship between Christianity and secular culture received at least five mutually exclusive types of answers. The variety of solutions proposed does not exclude the possibility of reaching general agreement on the level of fundamental issues. Painful misapprehensions arose on this level when more attention was paid to social-cultural factors generating apparent conflicts, than to the substantial content of the religious attitude toward scientific and human values.

When in today's philosophical anthropology there appear very often tones of despair and tragedy, when conditions of the human species are compared to the status of a cosmic gipsy or Sisyphus with a computerized stone-pushing system, the elements of creativity and openness that stem from religious premises keep us from aesthetically attractive philosophy of easy pessimism. The content of contemplative reflection upon transcendent values detracts our attention from comfortable narcissism and directs concern toward new domains that can enrich our existence. The enrichment presupposes acknowledgment of both scientific and human values. Their former critique from the standpoint of the religious world view resulted from the fact that in the atmosphere of transitory fascinations these two domains were regarded as sources of absolute imperishable values. From the perspective of the history of science we can assert categorically that which was supposed to be absolute and ultimate must be acknowledged as relative.

The acceptance of the religious Absolute permits us to assess the role of relative values and to develop an axiological theory that presupposes some absolute and unquestionable truths. In such a framework, intellectual creativity founded on religious values facilitates the building of a more human society. This society makes use of scientific discoveries in a manner substantially different from that of the well-known heroes of the Huxleyan ***Brave New World.*** The image of human society open toward transcendent religious values indicates a real chance to avoid the threat of consumer automatons of the XXI century. In this image, religious values appear as the enemy number one of the closed and narcissistically self-centered society. In the religious framework, the idols of our day, technology, production and institutional structures possess finite and relative value. Instead of fighting against these idols, contemporary successors of the prophets should look for means to subordinate the relative to the Absolute and to reach the goal of personal fulfillment.

10

In Defence of Scientific Reason

A New Aspirin of the Masses?

Former mystics, after experiencing disappointments and failures, very often became radical skeptics. The discovery of the illusory character of previously held beliefs in a new brave world, based upon scientific values, resulted in the skeptical denial of any objective value of scientific research. Unrealized hopes generated bitterness. Colorful dreams ended in frustrating black-and-white pictures of science in which human values must be necessarily demolished. Such a picture inspires various critical opinions in which we are informed that nasty science "destroys, dehumanizes, impoverishes, strips," etc. The authors of rhetorical compositions repeating the quoted verbs very often also repeat the basic objection that science provides neither values nor axiological suggestions needed for building a world adequate for human aspirations and ideals. In the same manner, however, we could attack music because it cannot satisfy our physiological hunger, or we could criticize sunrises since they disturb our sleep, which is so important for physical health.

If we correctly understand the role and tasks of music, we cannot complain that its beauty has not been a uiversal means of satisfying every form of human desire. In the same way, if we appropriately conceive the role and scope of science, nobody should regard scientific achievements as a source of universal happiness that is to satisfy every form of human dreams. If science is not thought of as an opium for the enlightened, but at most as an aspirin for pedestrians, nobody should expect either its universal application or miraculous effects. If appropriately applied, the aspirin can yield some positive results; if not, it can harm our organism. It is not the aspirin that should be blamed, but its careless users. The same remains true in the critique of science. It is not science's fault that it had been inappropriately used by ideological preachers who proclaimed the forthcoming of the new scientific age. The manifest exaggerations contained in their prophecies justify neither the total rejection of science nor the calling into question of its objective value.

The Christian attitude towards human intellectual efforts necessarily implies recognition of the objective value of the natural sciences. Regardless of its possible practical functions, the so-called scientific truth possesses an objective value because of the very nature of truth. One should admit that the growth of scientific theories is not a simple process of accumulation of discoveries. In contemporary science we find neither ether nor flogiston, so important for physics and chemistry of the past. Their presence in the science of the 18th century inspired theological debates that enabled the finding of better theories which express the truth of the physical world in a more adequate way than did the former theories. In my opinion, the scientific adventure should be conceived as a series of approximating theories which successively enlargen our knowledge of reality. Even if a scientist, for methodological reasons, eliminates the notion of truth of his vocabulary and restrains himself to technical approach which depends on an adopted standard of scientific research, the very accordance between research procedures and corresponding methodological rules reflects the objective directing of science towards truth. This feature, indeed, has been the essential element which differentiates scientific theories from their pseudo-scientific counterparts found in the works of Lysenko or Velikovsky.

In the Christian hierarchy of values, any efforts to discover objective truth appear as being related to Christ, Who named himself the Way, The Truth, and the Life. Scientific research, based on objective principles of scholarly investigations, are therefore objectively related to the truth which, according to Christ himself, is to liberate mankind (John 8:32). Certainly, very often newly discovered truths are overused to enslave the people and result in new prejudices, illusions and half-truths. The social abuse of science does not substantiate, however, denying its objective value. The so-called human values were also often misused in spectacular propaganda practiced by self-appointed defenders of humanism. The same might be said of religious values, when interdenominational rivalry and sociological factors turned out to be more important than the basic principles of love and tolerance.

All values, not just scientific ones, are sometimes inappropriately used in social practice. The practice itself, however, does not provide the ultimate criterion for the objectivity of values. Unrecognized truths, for instance, the Pythagorean theorem or evolutionary principles, constituted the objective truth even in the period when nobody knew of their existence. Bach's sonatas keep their objective aesthetic value even if a given group of their listeners loves nothing but jazz. Paintings of Rembrandt would remain masterpieces even if a rich and primitive owner used them merely to manifest his snobbish style. For the same reason, the well-known negative phenomena that occur in a scientific-technological society cannot justify the critical rejection of science. The biblical principle "by their fruits you shall know them," when properly understood, does not allow easy generalizations. If the fruits are rotten because they were badly stored, one should not blame the fruit tree, but rather an irresponsible keeper.

In the Christian hierarchy of value, the objective content of science has been much more important than its possible consequences, good or bad. For this reason, any form of genuine scientific reflection possesses an objective value, regardless of its possible practical applications. I would like to develop an analogy that exists between the attitude of a scientist working on new theories and the attitude of Mary kneeling in contemplative fascination at Christ's feet in Bethany. The contemplative insight of Mary is approved by Jesus, independently of the troubles of pragmatic Martha. Both the spiritual heritage of contemplative tradition and the scientific reflection aimed at the truth represent objective values that are essential for richness of human culture.

During three centuries of the coexistence of modern science and the Christian culture, very often bilateral simplifications hampered an objective appraisal of common goals. Isolated conflicts were often treated as manifestations of an essential conflict between religion and science. Mutual prejudices that flourished in both the narrow fundamentalism and the lyrical mythology of positivism facilitated mistrust between scientists and theologians. Distressing results of the mutual lack of confidence may be found, for example, in many 18th century works of Christian philosophers of nature. In their expositions, the most basic issues were avoided for diplomatic reasons in order to counteract possible suspicions that the author supported either risky new physics or the unorthodox philosophy of pantheism. Such intellectual diplomacy led to a practical neglect of the biblical image of God who permeates his creation and whose presence may be discovered in the beauty of nature. The thesis of divine immanence in the world, though theoretically unquestioned, was in practice reduced to vague formulae devoid of exact content.

The God that was required in the 18th century philosophy of nature was a God of scientific gaps. His existence was needed to eliminate explanatory difficulties from scientific theories and to justify the very fact of the existence of the world machine. Philosophers who adopted intellectual diplomacy restricted themselves to examining the doctrine of divine transcendence. For scientists, the doctrine of divine immanence in nature became useless since their explorations were focused upon observable quantitative aspects of physical reality and not upon its ultimate ontological ground. In such a context, the exciting issue of God's immanence in the world remained no man's land for a long period. Some theologians, keeping in mind painful recollections of past clashes between theology and science, tried to avoid problems which would result in new conflicts. In a similar way, many scientists who accepted the positivist theory of science tried to avoid ontological questions regarding the latter as trivial and meaningless.

The separation between scientific and religious culture became a natural consequence of adopting two radically different sets of cognitive means that ascertain reliability of our knowledge.

In various intellectual circles, the separation was regarded not only as the fact, but also as a necessity. The authors who believed religion to be an opium of the masses and the relic of a pre-intellectual stage in the growth of the human species never asked how to integrate scientific and religious values within one human culture. Correspondingly, some representatives of natural theology, disappointed that the growth of science put an end to many theological simplifications, treated new physical theories with distrust and arrogance, characteristic of ignorant people who are proud of their ignorance. As a result, "the whole earth", that for the Old Testament writers appeared as "filled by the spirit of the Lord," (Wisdom 1:7) for our generation appears mainly as a provincial planet surrounded by an aura of sophisticated scientific theories which often turn out to be mutually incompatible even if they belong to neighboring domains of scientific research.

In the chaos of competing interpretations and in the epistemological confusion of post-positivistic science, we should return to the basic questions formulated by the author of Psalm 139 who asked God:

> Where can I go from your spirit?
> from your presence where can I flee?
> If I go up to the heavens, you are there;
> if I sink to the nether world, you are present there.
> If I take the wings of the dawn,
> if I set at the farthest limits of the sea,
> Even there your hand shall guide me,
> and your right hand hold me fast.
>
> Psalm 139:7 - 10.

In the western sea of contemporary anti-scientific critiques and in the Sheol of naîve expeditions in occultism, we must keep the insight of the Psalmist and attempt to discover the presence of the divine hand in the fascinating book of Creation. The drama of our split cultures results to a great extent from the fact that, despite the good counsel given by Galileo, the scientific book of creation had many inattentive readers who examined its content selectively.

In Him We Live

In his impressive depiction of a lantern that moves along the night and brings sublime light into the heart of darkness, Gerard M. Hopkins provides a beautiful allegorical account of our intellectual condition. Night and distance may be approached without frustrating anxiety if, in the style of the author of *The Lantern Out of Doors*, we try to discover the subtle presence of divine light in the darkness of our intellectual divisions and existential frustrations. After the discovery of the subtle presence of the immanent Poet of the World our "darkness itself is not dark, and night shines as the day ." (Psalm 139:12). The horizon of our human existence appears in entirely new forms.

In the history of the intellectual development of mankind there were periods of optimistic belief in absolute brightness, the brigtness that was supposed to be generated by science, reason or historical necessity. In the variety of religious attitudes, there were also offers of easy escape in anti-intellectual isolation, based on distrust of both scientific and human values. When many former delusions are dispelled, we can no longer look for a single radical solution in which all disturbing problems are solved automatically. Similar solutions are nothing but phantasy products of ideological visionaries who would like to make mankind happy with their "only true" conceptions. The cultivation of similar dreams about a universal panacea that eradicates all negative phenomena from our society would be self-deceptive in the present epoch of prevailing pluralism, when universal "only true" standards are called into question even in the foundations of mathematics and physics.

The philosophy of *panentheism* and the doctrine of divine immanence in nature should not, therefore, be regarded as a universal remedy able to heal all sore places in our technological culture. Their functions are much more modest and we can compare their role to the role of the night lantern from Hopkins' poem. In the darkness of dehumanizing influences and at the frontiers of anti-scientific arguments there still remains a subtle light that strengthens our hope of overcoming the darkness. The light does not guarantee a happy end of our intellectual journey; the alleged happy end, announced in the past, turned out after all to be neither happy nor an end. Calling to mind former great promises and great disillusions, we may regard, nevertheless, the doctrine of panentheism as a happy means of integration of our culture. The means both remains psychologically akin to the human religious experience of the inward divine presence and seems to provide the most mature form of natural theology in which God is regarded as transcendent as well as immanent.

The apprehension of two complementary aspects of the divine reality in panentheism permits the construction of theological models (1) that satisfy both our need for local ordering and our aesthetic feelings. God, discovered at the heart of the universe, appears as God both of our rational explorations and of existential anxiety. He remains akin to the philosophical thought of Augustine, the contemplative insight of Francis of Assisi and the theological reflection of the epistle to the Colossians where the whole universe reveals the glory of the Lord, Who "is before all else that is. In him everything continues in being. ... It pleased God to make absolute fullness reside in him and, by means of him, to reconcile everything in his person, both on the earth and in heavens..." (Col 1: 17, 19f). This cosmic integration of all things, when consistently pursued in our philosophical analysis and axiological assessment, creates a real opportunity to overcome the present phenomenon of separated cultures. The awareness of the religious dimension of life requires constant rememberance of the Pauline words:"In him we live, and move and have our being." (Acts 17:28).

The sphere of scientific and human culture, when seen in the light of the words spoken by St. Paul to the Athenians on Mars Hill, turns out to be a sphere of hidden divinity which embraces the complex variety of our human acts. The variety contains numerous types of art and multifarious philosophic systems. Its inward directing toward beauty and truth reveals the glory of God Who "is all in all, ... greater is than all his works." (Sirach 43: 28f).

God's integrating presence that permeates both nature and culture constitutes a source of inspiration for our integrative efforts. God who, according to Pope's *Essay on Man*, "fills, ... bounds, connects and equals all" appears as a deep reality underlying the surface of fascinating data. Well-known attempts to stop on the surface level, proposed either in the present worship of nature or in traditional pantheism, are nothing but a philosophy of easy resignation. If we consistently try to wade through the surface of fascinating experiences, we discover, on the trail of Blake and Hopkins, "the dearest freshness deep down things". Its irradiating presence may be found in the rational exploration of the world, even "though the last light off the black West went" and though new stylish prophets proclaim the death of man and the end of human culture.

11

Love of Life or Fondness for Beauty

You throw the sand against the wind
And the wind blows it back again.
And every sand becomes a gem
Reflected in the beams divine.
William Blake

It was Edward O. Wilson who introduced the term biophilia to denote the tendency of the human species to focus its perception on life and life-like processes.(1) Even if one does not share Wilson's view that the tendency is genetically conditioned, it is impossible to deny that our contact with nature and the experience of its beauty provide strong feelings of exceptional cosmic unity. We feel disappointed when a bunch of flowers, that attracted our attention from a distance, turns out to be merely a plastic artifact. We enjoy the scenery of an autumn landscape of a valley with scattered trees. We contemplate the miraculous poetry of their colorful leaves. This experience of the beauty of life belongs to our strongest aesthetic feelings. The biophiletic dimension of human existence discloses a very important level of the relationship between the human species and the rest of the animate world. I am not convinced, however, that, as it has been suggested by representatives of sociobiology, human biophilia can be ultimately explained by taking into account the evolutionary history of our species and the genetic legacy inherited from our ancestors.

Sociobiological explanation remains attractive because of its simplicity and references to genetic links. The same merits, however, could have been ascribed to the mechanistic interpretation of nature developed in the 19th century. Neither its formal beauty nor taking into account the chain of deterministic links between explored phenomena sufficed to defend the simple vision of the world-machine. In my opinion the same may be said of the sociobiological explanation of biophilia. It is hardly enough to assume that homo sapiens is a poetic species sensitive to manifestations of life-like processes. The conjecture that this sensitivity was encoded in our genes remains at least untestable, because we cannot investigate empirically human aesthetics in the same manner as we explore anatomical traits of Drosophila. Sociobiological reference to genetic links thus remains as attractive as mechanistic attempts at a reductionist interpretation of nature. The attractiveness and simplicity of the interpretations proposed does not imply their truth.

One important element that should be taken into consideration in our analysis of biophilia is that the aesthetic fascinations experienced by us in contemplating the beauty of life resemble in many respects aesthetic feelings related to peculiar lifeless phenomena. When we enjoy watching sunsets, facing foggy mornings in the mountains, as well as when we contemplate regular structures of snowflakes and observe a lake's surface in moonlight, our psychic excitement seems to be akin to the biophiletic enchantment. Certainly, one could argue that water and sunlight, snow storms and the poetry of nightly feelings have been particularly connected with the phenomenon of life. Our feeling of their fantastic beauty is, therefore, nothing but a manifestation of human biophilia. If this argument, however, is consistently developed, one should expect biophiletic fascinations in the experience of all phenomena that are strictly connected with the genesis of life. For example, in the process of cosmic evolution an appropriate amount of stellar hydrogen was needed to obtain physical conditions necessary for carbon-based life. As has been shown in the works on the so-called *Cosmological Anthropic Principle* (2), a special rate of cosmic expansion and an extremely low density of matter were required to permit the process of carbon-based biogenesis. In spite of the amazing character of the discovered cosmic connections that enable the existence of life, one can hardly argue that we have a special emotional involvement into measurement of matter density or into the exploration of chemical properties of hydrogen. Consequently, one cannot substantiate the thesis claiming our biophiletic inclinations are present in all processes related, at least indirectly, to the phenomenon of life.

Our fascination with the beauty of nature deals not only with life-like processes. It seems to presuppose a more profound layer of being than is suggested by Wilson in his theory of biophilia. What is important in our strong aesthetic experience of the beauty of the world is that the domain of this experience embodies living creatures as well as lifeless structures. It is easier to contemplate the beauty of an autumn pasture full of birds than to recognize the amazing beauty of Einstein's field equations. These two distant fields of experience contain the same element of enchantment by a mysterious harmony that manifests itself both in the subtle poetry of everyday situations and in the sophisticated simplicity of mathematical relations. The recognition of the reality of the former would result in a half-truth if we completely ignore the amazing psychological force of the latter type of experience. It has been the very experience that fascinated the Hellenic admirers of logos and inspired Kepler's comments on symmetry of snowflakes; its expression may be found in the Old Testament's Book of Wisdom and in Plato's ontological investigations.

It was Plato who in the *Symposium* and in the *Hippias Major* introduced aesthetic categories to describe the ultimate ontic structure of reality. The Absolute Beauty from which all beautiful things derive their beauty constitutes in Plato's ontology "the wide ocean ... of lovely and majestic forms of beauty." Philosophical reflection, when consistently developed, becomes what I call kalophilia, the fondness of beauty.(3) Our intellectual efforts, according to this philosophy, aim at exploring the hidden beauty which may be found on our horizon of the observed play of parameters. Their transient existence points to a different reality which appears "indestructible, neither subject to increase nor decay, ... but eternally self subsistent... . All other things are beautiful through a participation of it, with this condition, that although they are subject to production and decay, it never becomes more or less, nor endures any change."(4)

Metaphysical analysis of the aesthetic aspects of being seemed a risky enterprise at the time when positivism dominated philosophy. The key objection that metaphysics contains nothing but expressive poetry **restrained** many authors from developing the doctrine of classical philosophy in which notions of beauty and being are convertible. It was Whitehead who dared to introduce aesthetic categories into his metaphysics at the time of the positivist disapproval of the poetic attitude in philosophy. His vision of God as the Poet of the World reestablishes the aesthetic categories in natural theology.

The Whiteheadian immanent God, cooperating with us in our efforts aimed at the beauty of harmonious existence, resembles in many respects the divine Wisdom described in the Old Testament by the author of the Book of Proverbs. The Divine Wisdom shows its presence in a variety of natural processes. Its "playing on the surface of earth" (Proverbs 8:31) reveals the harmony of the divine Logos that permeates the sphere of human experience. Our experience of moral beauty and our discovery of the aesthetic dimension of nature has by its nature a religious component. Its significance can hardly be overestimated when divine Wisdom describes its immanence in the human world in terms of God's delight of his communion with us. The "close to us" God of the Book of Proverbs has not too much in common with the Aristotelian Immovable Mover. His presence may be discovered not in the empyrean heavens but in the depths of our being and in the play of worldy shadows.

The fondness of beauty implies in its latent layers the implicit fondness of God, even if we do not recognize explicitly His existence. The immanent divine Logos discloses its subtle presence in our moral ideals, difficult ethical choices and aesthetic fascinations. It is the same Logos which unifies the divine Absolute with the ephemeral elements of our existence; as we read in the Book of Proverbs:

> When he established the heavens I was there,
> when he marked out the vault
> over the face of the deep;
> When he made firm the skies above,

when he fixed fast the foundations of the
earth;
When he set for the sea its limit, so that
the waters should not transgress his command;
Then was I beside him as his craftsman,
and I was his delight day by day, playing
before him all the while, playing
on the surface of his earth;
and I found delight in the sons of men.
Proverbs, 8:27 - 31.

The insight combining aesthetic feelings and rational reflection, contemplative openness to the hidden depths of philosophical amazement at prosaic situations may be found relatively easily in the works of contemporary mystical authors who search for invisible divine beauty hidden in nature. It is the very mystical kalophilia, the fondness of beauty rooted in a biblical vision of nature, that strikes us in Thomas Merton's descriptions contained in his *Sign of Jonas*. Merton's poetry of nature can be reduced neither to biophiletic fondness of life nor to natural aestheticism. In his perspective, various trends and opposing tendencies became unified after disclosing their transcendent beauty and indicating their ontic origins in the hidden ultimate source of Beauty. Rational elements contained in such an approach prevent it from ending in purely subjective aestheticism.

The infinite being of the immanent God provides the richness of content that requires a permanent freshness of our insight and focuses the insight upon new unknown aspects of the explored reality. In Merton's description of such an exploration we contemplate "the blue elm tree at hand and the light blue hills in the distance... Delightful darkness, delightful sun, shining on a world... The hills are as pure as jade in the distance. God is in His transparent world, but He is too sacred to be mentioned, too holy to be observed. I sit in silence."(5)

The contemplative silence discloses the hidden depth of the world. Transparent nature reveals the immanent divine source of its beauty and integrates our feelings. This experience of cosmic unity and divine beauty seems to grow out of roots deeper than the biophiletic roots of the human species. Its depth reaches the ontic level of our being where the ultimate source of harmony can be found. The religious openness to this level was, for cultural or methodological reasons, often hampered in the past. Today we realize how naïve was the thesis of doctrinaire positivism claiming that "the world is all surface." Sharing Peter B. Medawar's critique of positivism as well as his assessment of science and reason, we may follow him in arguing that scientific explanations of our position and goals in the universe are necessary but not sufficient. (6) When we go beyond scientific theories and chains of syllogisms, we can discover the horizon of subtle mystery, hidden beauty, immanent divine presence. In this experience, a new reality is discovered disclosing a deeper world permeated by the Creator of all Beauty. When contemplating his presence we affirm that "we must be in love with him, not with ourselves or our schemes, we must find time to let our minds dwell on him. The beauty of holiness in the midst of this revolutionary world belongs to those who set the Lord always before their eyes."(7) Such an attitude cannot result from uncritical worship of "the" scientific method.(8) Neither can it stem from purely rational explanations of nature. We need both aesthetic sensitivity and contemplative insight to discover the hidden presence of divine holiness and to experience its fascinating beauty.

At the beginning of Western rational thought, Hellenic philosophers raised the ultimate questions of the arche, substance, basic principles of being. Philosophia, continuing the tradition of pre-Socratic wisdom, became an important element that unifies various cultures end epochs in the same search for rational explanation of the human world. Kalophilia, searching for a hidden aesthetic component of our existence, creates a new perspective of unity between cultures that were separated because of simplifying theories of knowledge.

Notes

NOTES TO INTRODUCTION

1. H. Adams, The Education of Henry Adams, (Boston: Haeughton Miffin, 1918) 380.

2. F. Klingender, Art and the Industrial Revolution, (London: Paladin, 1975) VII.

3. A. Maslow, The Farther Reaches of Human Nature, (New York: The Viking Press, 1966) 195.

4. Cf. M. Ruse, 'Evolutionary Epistemology: Can Sociobiology Help?' in Sociobiology and Epistemology, ed. J.H. Fetzer, (Dordrecht: D. Reidel, 1985) 255:"Refusal to accept unseen entities may lose 'you' the Nobel Prize."

5. Cf. A.N. Whitehead, Adventures of Ideas, (London: The Free Press, 1967) 295.

NOTES TO CHAPTER 1

1. N. Copernicus, Complete Works, vol. 2, (Warsaw: Polish Scientific Publishers, 1978) 5.

2. Cf. O. Pedersen, Galileo and the Council of Trent, (Citta del Vaticano: Specola Vaticana, 1983).

3. Galileo, Le Opere di Galileo Galilei, vol. 7, ed. A. Favaro, (Florence: G. Barbera, 1890-1909) 35, 129.

4. The most optimistic expression of Pope's approach to Newton's physics is contained in his famous text:

Nature and nature's laws lay hid in night;
God said: Let Newton be! and all was light.

5. A more detailed presentation of the intellectual legacy of the School of Chartres may be found in M.D. Chenu, 'Nature and Man at the School of Chartres in the Twelfth Century,' in The Evolution of Science, eds. G.S. Metraux, F. Crouzet, (New York: Mentor Books, 1963) 220 - 235.

6. J. Kepler, Harmonies of the World, in: Great Books of the Western World, ed. R.M. Hutchins, (Chicago: 1952) 1085.

NOTES TO CHAPTER 2

1. See, for instance, A. Naess, Einstein, Spinoza and God, in: The Old and New Questions in Physics, Cosmology and Theoretical Biology, (New York, 1983).

2. A general description of the phenomenon of neognosis may be found in R. Ruyer, La Gnose de Princeton, (Fayard, 1974).

3. Cf. Science et conscience. Les deux lectures de l'univers, (Stock et France Culture, 1980); L'Esprit et la science. Colloque des FES, (Albin Michel, 1983).

NOTES TO CHAPTER 3

1. C.P. Snow, The Two Cultures and the Scientific Revolution, (Cambridge: CUP, 1961).

2. Cf. C. Truesdell, An Idiot's Fugitive Essays on Science, Methods, Training, Circumstances, (Springer, 1984) 585f.

3. Cf. J. Passmore, Science and Its Critics, (London: Duckworth, 1978).

4. Copernicus, Complete Works, vol. 2, 5.

5. P.C.W. Davies, God and the New Physics, (New York: Simon and Schuster, 1983) IX.

NOTES TO CHAPTER 4

1. Cf. Ch. Castonguay, Meaning and Existence in Mathematics, (New York, 1972).

2. A. Einstein, Ideas and Opinions, (New York, 1976) 49.

3. H. Pagels, The Cosmic Code. Quantum Physics as the Language of Nature, (New York, 1973).

4. P. Medawar, Pluto's Republic, (Oxford: OUP, 1983) 242.

NOTES TO CHAPTER 5

1. C.W. Misner, K.S. Thorne, J.A. Wheeler, Gravitation, (San Francisco: Freeman, 1974) 1215.

2. A.N. Whitehead, Process and Reality. An Essay in Cosmology, (New York 1969) 408.

3. J.F. Cobbs, God and the World, (Philadelphia, 1976) 82.

4. A.R. Peacocke, Creation and the World of Science, (Oxford: OUP, 1979) 204.

5. J.V. Taylor, The Go-Between God, (Oxford: OUP 1972).

6. A popular presentation of Harthshorne's philosophy is contained in E. Peters, The Creative Advance, (Bethany Press, 1966).

NOTES TO CHAPTER 6

1. The most developed theory of God's co-suffering with mankind may be found in F. Varillon, La Souffrance de Dieu, Paris 1978. This book was awarded the Grand Prix Catholique.

2. See Ch. Hartshorne, Whitehead and Berdyaev: Is there Tragedy in God? "Journ. of Rel." 37 (1957) 83; A New Look at the Problem of Evil: Current Philosophical Issues, Springfield 1966.

NOTES TO CHAPTER 7

1. J.A.T. Robinson, Exploration into God, (SCM Press, 1967) 86.

2. Cf. H. Montefiore, The Probability of God, (London, 1985).

3. Cf., for instance, K. Rahner, H. Vorgrimler, Kleines Theologisches Woerterbuch, (Freiburg, 1961) 275.

NOTES TO CHAPTER 8

1. R. Dubois, A God Within, (New York, 1972).

2. T. Merton, The Waters of Siloe, (New York, 1962) 21.

3. T. Merton, Mystics and Zen Masters, (New York, 1967) 114.

4. K. Barth, The Knowledge of God and the Service of God, (Hodder, 1928) 5.

5. In the same Calvinist tradition to which Barth belonged, the role of the rational component of religious faith is emphasized, for instance, by Alvin Plantinga. A thorough analysis of his valuable contributions may be found in "Alvin Plantin-

ga", ed. J.E. Tomberlin, P. van Invagen, (Dordrecht: D. Reidel, 1985).

6. Bonaventure, Collationes in Hexamaeron, 12.5.

NOTES TO CHAPTER 9

1. E. Fromm, The Revolution of Hope. Toward a Humanized Technology, (New York: Bantam Books, 1968) 143.

2. Cf. R.D. Masters, 'Evolutionary Biology, Human Nature, and Knowledge,' in Sociobiology and Epistemology, ed. J.H. Fetzer, (Dordrecht: D. Reidel, 1985) 97, 110.

NOTES TO CHAPTER 10

1. The issue of theological models is discussed in detail by Ian G. Barbour in his Myths, Models, and Paradigms, (New York: Harper and Row, 1976) esp. chap. III and IV.

2. A Hopkins Reader, ed. J. Pick, (New York: Doubleday, 1966) 48.

NOTES TO CHAPTER 11

1. E.O. Wilson, Biophilia, (Cambridge, Mass.: Harv. Univ. Press, 1983).

2. See J.D. Barrow, F.J. Tipler, The Anthropic Cosmological Principle, (Oxford: OUP, 1986).

3. In Greek, "kalos" means beautiful; " philein", to love.

4. Plato, Symposium, 210e 1 - 212a 7.

5. T. Merton, The Sign of Jonas, (New York: Harcourt, 1953) 338.

6. P.B. Medawar, The Limits of Science, (New York: Harper and Row, 1985) 98, 101.

7. J.V. Taylor, The Go-Between, (Oxford: OUP, 1972) 242.

8. Cf. Medawar, The Limits of Science, 51: "there is indeed no such thing as 'the' scientific method. A scientist uses a very great variety of explanatory strategems, and although a scientist has a certain address to his problems, ... he uses no procedures of discovery that can be logically scripted."